Rapid Math Tricks & Tips

30 Days To Number Power

EDWARD H. JULIUS

John Wiley & Sons, Inc.
New York / Chichester / Brisbane / Toronto / Singapore

Published by John Wiley & Sons, Inc.

Library of Congress Cataloging-in-Publication Data:

Julius, Edward H., 1952–
Rapid math : Tricks & tips, 30 days to number power / Edward H. Julius.
p. cm.
ISBN 0-471-57563-1 (pbk.)
1. Arithmetic. I. Title.
QA111.J85 1992
513′.9–dc20 92-10638

Printed in the United States of America

10 9 8 7 6 5 4 3 2 1

To Marion, Marina, and Alexandra

Acknowledgments

I would like to gratefully acknowledge the following individuals, without whose input and generosity this book would have been written anyway (but it wouldn't have turned out nearly as well): Dr. Lyle Sladek, Debbie Weber, Susan Sunde, and especially my brother Ron.

I'd also like to thank Chris Smith for lending his cartooning skills, and Judith McCarthy and Steve Ross of John Wiley & Sons for believing in me.

Contents

For your enjoyment, a number of mathematical parlor tricks, curiosities, and potpourris have been interspersed throughout this book.

— Introduction —

I have often admired the mystical way of Pythagoras, and the secret magic of numbers —Sir Thomas Browne

Welcome to the fascinating world of rapid calculation! I am delighted to be your guide, and I promise to make the journey as interesting and enjoyable as possible.

Prepare yourself for an engaging self-study program that will forever change your approach to numbers. The more energy you focus on "number power," the greater the likelihood that you will master the techniques presented. Your ultimate reward will be their practical application.

For me, this book is the culmination of 30 years of learning, researching, and applying rapid calculation techniques, both in and out of the classroom. Some techniques are fairly well known in the world of mathematics. Others I have discovered on my own over the years. More important, I use most of these techniques regularly. With practice, so will you!

What Will I Get Out of This?

You might be thinking, "How will I benefit from this 'number-power' program?" Consider the following situations.

When was the last time you were at the supermarket, and you had only $25 with you? You're standing in line at the checkout counter, terrified that you've purchased more than $25 worth of groceries. Several other customers are looking on. You begin to shake. You start to think about which items to put back. You envision an embarrassing scene, followed by ejection from the store, a lengthy jail sentence, or worse!

Perhaps you are at a restaurant with some friends. You've offered to leave the tip, but discover that you've lost your calculator. You break out in a cold sweat and wonder, "What do I do now?" Your whole life begins to pass before your eyes. Your friends wait impatiently.

As you may have guessed, this number-power program was designed to put *you* in command, so that you will *never* have to be intimidated by numbers again. But wait—there's more! Whether you realize it or not, you will also develop a better feel for numbers and their interrelationships. You will develop more confidence in your math skills while your anxiety and fear of numbers disappear.

Your ability to process and store information will grow, as will your self-esteem. And think how you'll feel when you astound your friends and relatives with your newfound talents.

Most important, I know that you will find this program motivating, fascinating, and fun!

But Isn't This the Age of the Calculator?

You might wonder, "Why should I rack my brain when I can simply let my calculator do all the work?" Good question, but I think that I've got some good answers:

- Once you've mastered this number-power program, you'll never have to rack your brain again.
- You're not always allowed to use a calculator. When taking standardized tests, for example, you normally must rely upon (gasp!) pencil-and-paper calculations.
- Often, it is awkward or inappropriate to use a calculator. For example, if you are a teacher who performs frequent calculations on the board, you don't want to look like a total innumerate by referring to your calculator every 12 seconds.
- Sometimes you want to perform a calculation secretly. For example, if you think that a department store clerk has short-changed you, you won't have to give the appearance that you are challenging his or her

competence. (In fact, in some cultures it is considered an insult to inspect your change.) Instead, you will have the confidence to challenge the clerk and know that you're right. This can be especially useful in restaurants when you only have a few seconds to examine the check.

- Very often, mental calculations can be performed more quickly than with a calculator. This is especially true when all you need is an estimate.
- The only calculators that are convenient enough to carry around are so small that you can go blind trying to use them.

What Else Do I Have to Know about Rapid Math?

As you work through this number-power program, please keep in mind the following:

- You must do three things to ensure mastery of the tricks presented—practice, practice, practice! (Or is that how you get to Carnegie Hall?)
- The more tricks you master, the more often you'll be able to apply number power.
- A trick is beneficial only if it is not too cumbersome and is truly faster than the traditional method. It must also be practical. So the 17-step trick to calculate division by 483 did not make it into this book.
- Even if a trick saves only a picosecond (that's one trillionth of a second), it's worth using. Well, maybe just a split second.
- Because addition and multiplication are, for some strange reason, generally faster and easier to execute than subtraction and division, you will be guided in that direction whenever possible.
- You'll need to hide your calculator immediately! I'll pause for a moment while you do that.

But Do I Have What It Takes?

To become a number-power wizard, you don't have to be a descendant of Albert Einstein, work as a rocket scientist, or possess an advanced degree in differential calculus. All you need is a basic understanding of addition, subtraction, multiplication, and division. That's all.

Oh, I forgot—to solve the optional "Brain Builder" problems, you'll also need a basic understanding of decimals and fractions. It would also be very helpful to review the math concepts on page 8.

The most important requirement for this number-power program is the desire to learn rapid math. You obviously have that desire, because you were curious enough to buy this book.

I designed this number-power program to accommodate people of diverse backgrounds, ages, and abilities. Accordingly, I would be eternally grateful, or at least somewhat appreciative, if you would show patience and understanding if the pace is too slow or too fast for you.

I'm Ready. Where Do I Go from Here?

To get the most out of this program, first finish reading the introduction. If you don't normally read introductions (I never do), read this one anyway. (Now the $64,000 question: "How do I get you to read the introduction that tells you that you should read the introduction?" Catch-22.)

Anyway, after you've read the introduction, take the pre-test that follows. Then grade your answers—you will eventually "revisit" the pre-test. Next read the sections on brief review, symbols, terms, math tables, squares, and reasonableness. Then you will be ready to begin the 30-day program.

You'll see that each day's lesson consists of two tricks. For each trick, read the strategy and examples until you have a good feel for the rapid math concept and procedure. Don't overlook the "Number-Power Notes," for they provide special insight into the numbers. Then work through the practice exercises on your own, checking the answers at the back of the book.

If you want more than a basic ability in rapid calculation, work through the "Brain Builder" exercises and check your answers. As you proceed to each new rapid calculation trick, don't hesitate to use any trick you've already learned. Obviously, you can shorten or lengthen the program by altering the number of tricks you cover each day. However, I recommend against completing the program in one weekend by trying to learn 30 tricks per day!

At the end of each week, take the "Quick Quiz." If you have opted to cover only the basic program, simply skip the problems labeled "Brain Builders." When you have completed the work for day thirty, take the pre-test again and the final exam to see how much you've learned.

Finally, read the conclusion. This section contains some very important advice on how to apply these tricks daily, and on how to maintain your number-power abilities.

For your amusement, some dynamite "Parlor Tricks," "Mathematical Curiosities," and "Number Potpourris" have been spread throughout the book. For your convenience, a summary of the 60 number-power tricks has been provided for handy reference at the back of the book—you can use this as a quick refresher.

Reread the book from time to time—the goal is to have these tricks become automatic. With time, they will become second nature to you.

Teachers, Take Note!

I want to point out that *Rapid Math Tricks & Tips* is different from any other math book you've ever seen. It doesn't cover traditional math topics and it isn't geared toward any particular grade level. As such, you are probably wondering how and if it will fit into a math curriculum. You may be skeptical about how the whole idea of rapid calculation will be received by students.

First of all, this book was written with practically all age groups in mind. This was easy to accomplish because most people, even adults with math-related jobs, are never exposed to rapid calculation techniques. In fact, when I lecture on the topic, math teachers are usually among my most enthusiastic listeners! Accordingly, both the sixth-grader and the business executive can benefit from this number-power program.

At the very least, this book could be used to supplement a grade-level textbook. Just cover a dozen or two of the easiest tricks, and have your students do only the elementary exercises. For those students wishing to undertake additional topics on their own, the self-teaching nature of this book will enable them to continue on to the "Brain Builders" and to the more challenging tricks.

At its best this book constitutes the basis for an entire course in rapid calculation. As much as I have tried to illustrate each trick clearly and logically, there is no substitute for live, step-by-step instruction. Not only are you available to answer your students' questions each step of the way, but you can also monitor their progress as they work on the reinforcement problems in class. (With a little imagination and experimentation, you will find that this number-power program can fit into, and enhance, practically any math curriculum designed for almost any age.)

One final note—someone is bound to pose the question, "If these rapid calculation techniques are so terrific, why aren't they taught in school rather than the traditional techniques?" One answer is that a solid foundation in "traditional math" is a prerequisite to understanding the concepts of rapid math.

All Right, I'm Convinced. Now Can We Get This Show on the Road?

Yes, this is it—what you've been waiting for—the end of the introduction. So fasten your seat belt, put on your thinking cap, and get ready for a mathematics course unlike any other you've ever experienced!

Edward Julius

Pre-Test

How Many of These Calculations Can You Do in Two Minutes?

Work each calculation as quickly as you can, without using a calculator. Your time limit is two minutes, and you may work the problems in any order. When your time is up, move on to the next page and check your answers (don't look yet!). You will revisit these problems when you've completed the 60 rapid calculation techniques presented in this book.

1. $99 \times 85 =$
2. $700 \div 25 =$
3. $3.5 \times 110 =$
4. $4,600 \div 50 =$
5. $1.9 \times 210 =$
6. $425 - 387 =$
7. $31 \times 31 =$
8. $7 + 24 + 343 + 50 =$
9. $22 \times 18 =$
10. $31.5 \div 3.5 =$
11. $1 + 2 + 3 + 4 + 5 + 6$
    $+7 + 8 + 9 + 10 =$
12. $120 \div 1.5 =$
13. $65 \times 65 =$
14. $74 \times 101 =$
15. $163 - 128 =$
16. $109 \times 104 =$

Answers to Pre-Test

1.	$99 \times 85 = 8{,}415$	9.	$22 \times 18 = 396$
2.	$700 \div 25 = 28$	10.	$31.5 \div 3.5 = 9$
3.	$3.5 \times 110 = 385$	11.	$1 + 2 \ldots + 10 = 55$
4.	$4{,}600 \div 50 = 92$	12.	$120 \div 1.5 = 80$
5.	$1.9 \times 210 = 399$	13.	$65 \times 65 = 4{,}225$
6.	$425 - 387 = 38$	14.	$74 \times 101 = 7{,}474$
7.	$31 \times 31 = 961$	15.	$163 - 128 = 35$
8.	$7 + 24 + 343 + 50 = 424$	16.	$109 \times 104 = 11{,}336$

On the average, my college students are able to complete only a handful of these problems correctly in two minutes. Considering that most people perform calculations the traditional way, solving a handful in two minutes isn't bad. However, by the time you have finished this 30-day program, you should be able to complete most or all of them correctly in less than two minutes, as there is at least one number-power trick for each problem. You will eventually be able to identify the relevant trick immediately, and then perform the calculation in less time than you thought possible!

Brief Review of Some Basic Math Concepts

1. Subtraction is the inverse of addition:
   A. If $a - b = c$, then $c + b = a$
   B. So if $10 - 3 = 7$, then $7 + 3 = 10$
2. Division is the inverse of multiplication:
   A. If $a \div b = c$, then $c \times b = a$
   B. So if $45 \div 9 = 5$, then $5 \times 9 = 45$
3. Addition and multiplication follow the commutative principle:
   A. $a + b = b + a$, so $8 + 3 = 3 + 8$
   B. $a \times b = b \times a$, so $8 \times 3 = 3 \times 8$
4. Affixing zeroes to the left of a number or to the right of a decimal point will not change the number:
   A. So $84 = 084 = 0084 = 84.0 = 84.00$, and so on.
   B. Also, $26.9 = 026.9 = 26.90 = 26.900$, and so on.
   C. However, note: 84 does not equal 804, 4.7 does not equal 4.07, and 5.3 does not equal 50.3
5. To multiply a number by 10, 100, and so forth, simply affix the appropriate number of zeroes to the right of the number, or simply move the decimal point the appropriate number of places to the right:
   A. So $26 \times 10 = 260$, and $95 \times 100 = 9,500$
   B. Also, $8.17 \times 10 = 81.7$, and $3.14 \times 100 = 314$
6. To divide a number by 10, 100, and so forth, simply move the decimal point of the number the appropriate number of decimal places to the left:
   A. So $34 \div 10 = 3.4$, and $691.5 \div 100 = 6.915$
7. When dividing, you may cancel an equal number of "right-hand" zeroes:
   A. So $600 \div 30 = 60 \div 3$, and $8,000 \div 200 = 80 \div 2$

Symbols, Terms, and Tables

Symbols

1. "+" means "plus" or "add"
2. "−" means "minus" or "subtract"
3. "×" means "times" or "multiplied by"
4. "÷" means "divided by"
5. "=" means "equals"
6. "≈" means "approximately equals"
7. "n^2" means "a certain quantity squared"

Terms

1. AFFIX: To attach a digit or digits to the right of an existing number. For example, to multiply 37 by 100, you merely affix two zeroes to the right of the 37, producing an answer of 3,700.
2. APPROXIMATE RECIPROCALS: As used in this book, two numbers that will produce approximately 10, 100, or the like when multiplied together. For example, 9 and 11 are approximate reciprocals because their product is 99 (approximately 100).
3. CONVERSION TRICK: As used in this book, multiplication or division of two numbers in decimal form by using a number in fraction form (or vice versa). For example, to divide a number by 1.5, simply multiply that number by ⅔.
4. DIGIT: A single component of a number, such as the ones-place digit. For example, there are three digits in the number 371.
5. INSERT: To place a digit or decimal point in between digits of an existing number. For example, inserting a decimal point in the middle of 64 would produce the number 6.4.
6. INTERMEDIARY FIGURE: A temporary sum, product, difference, or quotient that is obtained when you have completed a portion of a calculation.
7. NUMBER: An entire numerical expression, such as 847 or 5,203.
8. NUMBER POWER: Rapid calculation; also known as rapid math or mental math.
9. RECIPROCALS: As used in this book, two numbers that will produce 10, 100, or the like when multiplied together. For example, 5 and 2 are reciprocals because their product is 10. The numbers 4 and 25 are reciprocals as well, because their product is 100. (In the world of mathematics, however, reciprocals are two numbers whose product is 1.)
10. SQUARE: A number multiplied by itself. For example, the square of 12 (or 12^2) is 12×12, or 144.
11. TEST OF REASONABLENESS (T of R): An examination of an intermediary figure or answer to determine merely whether it is "in the ballpark." See page 11 for a comprehensive explanation.

Operational Terms

```
  39 ← Addend        483 ← Minuend
+ 22 ← Addend      − 291 ← Subtrahend
----               -----
  61 ← Sum           192 ← Difference
```

$$\begin{array}{r l}
17 & \leftarrow \text{Multiplicand} \\
\times\ 56 & \leftarrow \text{Multiplier} \\
\hline
952 & \leftarrow \text{Product}
\end{array}$$

$$\begin{array}{ccccc}
\text{Dividend} & & \text{Divisor} & & \text{Quotient} \\
\downarrow & & \downarrow & & \downarrow \\
891 & \div & 33 & = & 27
\end{array}$$

Addition/Subtraction Table

+	1	2	3	4	5	6	7	8	9	10	11	12
1	2	3	4	5	6	7	8	9	10	11	12	13
2	3	4	5	6	7	8	9	10	11	12	13	14
3	4	5	6	7	8	9	10	11	12	13	14	15
4	5	6	7	8	9	10	11	12	13	14	15	16
5	6	7	8	9	10	11	12	13	14	15	16	17
6	7	8	9	10	11	12	13	14	15	16	17	18
7	8	9	10	11	12	13	14	15	16	17	18	19
8	9	10	11	12	13	14	15	16	17	18	19	20
9	10	11	12	13	14	15	16	17	18	19	20	21
10	11	12	13	14	15	16	17	18	19	20	21	22
11	12	13	14	15	16	17	18	19	20	21	22	23
12	13	14	15	16	17	18	19	20	21	22	23	24

Note: Any number plus zero equals that number. To use this table to subtract, find the sum in the body of the table and read up or to the left to determine the minuend and subtrahend. For example, 19 − 11 = 8.

Multiplication/Division Table

×	1	2	3	4	5	6	7	8	9	10	11	12
1	1	2	3	4	5	6	7	8	9	10	11	12
2	2	4	6	8	10	12	14	16	18	20	22	24
3	3	6	9	12	15	18	21	24	27	30	33	36
4	4	8	12	16	20	24	28	32	36	40	44	48
5	5	10	15	20	25	30	35	40	45	50	55	60
6	6	12	18	24	30	36	42	48	54	60	66	72
7	7	14	21	28	35	42	49	56	63	70	77	84
8	8	16	24	32	40	48	56	64	72	80	88	96
9	9	18	27	36	45	54	63	72	81	90	99	108
10	10	20	30	40	50	60	70	80	90	100	110	120
11	11	22	33	44	55	66	77	88	99	110	121	132
12	12	24	36	48	60	72	84	96	108	120	132	144

Note: Any number times zero equals zero. To use this table to divide, find the product in the body of the table and read up or to the left to determine the divisor and quotient. For example, 88 ÷ 11 = 8.

The Squares of the Numbers 1 Through 20

$1^2 = 1$	$6^2 = 36$	$11^2 = 121$	$16^2 = 256$
$2^2 = 4$	$7^2 = 49$	$12^2 = 144$	$17^2 = 289$
$3^2 = 9$	$8^2 = 64$	$13^2 = 169$	$18^2 = 324$
$4^2 = 16$	$9^2 = 81$	$14^2 = 196$	$19^2 = 361$
$5^2 = 25$	$10^2 = 100$	$15^2 = 225$	$20^2 = 400$

Table of Equivalencies

1/100	= 0.01	= 1%	3/8	= 0.375	= 37½%
1/50	= 0.02	= 2%	2/5	= 0.4	= 40%
1/40	= 0.025	= 2½%	1/2	= 0.5	= 50%
1/25	= 0.04	= 4%	3/5	= 0.6	= 60%
1/20	= 0.05	= 5%	5/8	= 0.625	= 62½%
1/10	= 0.1	= 10%	2/3	= $0.66\overline{6}$	= 66⅔%
1/9	= $0.11\overline{1}$	= 11⅑%	3/4	= 0.75	= 75%
1/8	= 0.125	= 12½%	4/5	= 0.8	= 80%
1/5	= 0.2	= 20%	7/8	= 0.875	= 87½%
1/4	= 0.25	= 25%	x/x	= 1.0	= 100%
1/3	= $0.33\overline{3}$	= 33⅓%			

The Test of Reasonableness

In general, applying a test of reasonableness (T of R) to an answer means looking at it in relation to the numbers operated upon to determine if it's "in the ballpark." Put in simple terms, you look at the answer to see if it makes sense. For example, if you determine 10 percent of $75 to be $750 (which could happen if you use a calculator and forget to press the percent key), you should immediately notice that something is very wrong.

The majority of the number-power tricks in this book involve multiplication and division. As you will learn very shortly, you should ignore decimal points and zeroes when starting these calculations. When you've completed the calculation, you will need to determine if and where to insert or affix a decimal point or zeroes based upon "what looks right."

For example, you will soon learn that to divide 13 by 5, you first double the 13 to arrive at an intermediary figure of 26. You then notice, by thinking about the division of 13 by 5, that 26 is much too large to be the answer. By inserting a decimal point between the 2 and the 6, you arrive at a number (2.6) that seems to fit the bill. With practice, you will be able to apply a test of reasonableness with ease.

A test of reasonableness can also be performed for addition and subtraction. For example, if you add 86 to 94 and arrive at 210 as an answer, you should know immediately that you've miscalculated because two numbers under 100 cannot total more than 200!

There are many more ways to apply a test of reasonableness, such as rounding to perform a quick estimate of the answer and comparing it with the intermediary figure you've obtained. To test your ability to perform the test of reasonableness, take the brief multiple-choice quiz below. **Do not** perform the calculations, except to "eyeball" them to get a feel for where the answer should fall. Then look at the choices and determine which one makes the most sense.

(1) $120 \div 8 =$
a. 1.5
b. 0.15
c. 150
d. 15

(2) $16 \times 22 =$
a. 352
b. 35.2
c. 3,520
d. 35,200

(3) $7,200 \div 960 =$
a. 750
b. 75
c. 0.75
d. 7.5

(4) $470 \times 6.2 =$
a. 291.4
b. 2,914
c. 29,140
d. 29.14

(5) $187 + 398 =$
a. 485
b. 585
c. 685
d. 785

(6) $452 - 255 =$
a. 197
b. 397
c. 97
d. 297

(Answers: 1.d, 2.a, 3.d, 4.b, 5.b, 6.a)

Week 1 Multiplication and Division I

Before you begin the program with today's techniques, make sure you understand the basic math concepts reviewed previously. In particular, you should be able to multiply and divide by 10, 100, and so forth, as illustrated in items 5 and 6 on page 8. Today's tricks are very important because they will be used as building blocks for many of the later tricks.

Trick 1: Multiplying and Dividing with Zeroes

Strategy: The first step to performing rapid multiplication or division is to **disregard any zeroes** comprising the right-hand portion of the numbers. For example, $1,200 \times 50$ should be viewed as a 12×5 computation. Then, to complete the problem apply a "test of reasonableness." That is, ask yourself, "How many zeroes will produce an answer that makes sense?" In this case, it would seem logical to affix three zeroes to the intermediary product of 60 (12×5) to produce an answer of 60,000. Remember that this rule applies to rapid multiplication and division but does not apply to rapid addition and subtraction, which follow a different set of rules. Let's go over some examples, step by step.

Example #1
30 × 70

Step 1. Disregard the zeroes and think, "3×7."

Step 2. Multiply: $3 \times 7 = 21$ (intermediary product).

Step 3. Apply test of reasonableness (T of R): Because two zeroes were disregarded initially, affix two zeroes to the intermediary product, producing the answer 2,100.

Thought Process Summary

$$\begin{array}{r} 30 \\ \times\ 70 \\ \hline \end{array} \rightarrow \begin{array}{r} 3 \\ \times\ 7 \\ \hline 21 \end{array} \rightarrow 2{,}100$$

Example #2

4,800 ÷ 120

Step 1. Disregard the zeroes and think, "48 ÷ 12."

Step 2. Divide: 48 ÷ 12 = 4 (intermediary quotient).

Step 3. Apply T of R: As explained previously, you may cancel an equal number of right-hand zeroes when dividing. Therefore, the problem becomes 480 ÷ 12. We know that 48 ÷ 12 = 4, so 480 ÷ 12 must equal 40 (the answer).

Thought Process Summary

$$4{,}800 \div 120 \rightarrow 48 \div 12 = 4 \rightarrow 40$$

Example #3

4,500 ÷ 900

Step 1. Disregard the zeroes and think, "45 ÷ 9."

Step 2. Divide: 45 ÷ 9 = 5. (This is the answer, since an *equal* number of zeroes from the original dividend and divisor were disregarded.)

Thought Process Summary

$$4{,}500 \div 900 \rightarrow 45 \div 9 = 5$$

Number-Power Note: When you have a number such as 800.6, the two zeroes would not be disregarded, because they do not comprise the entire right-hand portion of the number.

Practice Exercises

Here are some exercises to try—remember to forget about the zeroes when starting the calculation.

1. 40 × 7 =
2. 6 × 800 =
3. 500 × 30 =
4. 60 × 900 =
5. 70 × 120 =
6. 15 × 150 =
7. 400 × 50 =
8. 24 × 400 =
9. 3,600 ÷ 900 =
10. 5,600 ÷ 7 =
11. 5,200 ÷ 130 =
12. 800 ÷ 16 =
13. 42,000 ÷ 60 =
14. 1,800 ÷ 90 =

(See solutions on page 199)

Number Potpourri #1

The number "one googol," which is 10,000,000,000,000,000,000,000, 000,000,000,000,000,000,000,000,000,000,000,000,000,000,000, 000,000,000,000,000,000,000,000,000,000, can be easily written but couldn't be counted to in a million lifetimes. It is a number so large that it is estimated that there aren't even that many electrons in the universe. However, the number "one googolplex," which is a one followed by a googol zeroes, not only couldn't be counted to in a million lifetimes, it couldn't even be written in that same time span (except in scientific notation)!

Trick 2: Multiplying and Dividing with Decimal Points

Strategy: This is the second of two techniques to be used in conjunction with many of the later tricks. It is similar to Trick 1 and comes in handy when performing dollars and cents calculations. When you begin a rapid calculation, **disregard any decimal points**. For example, 2.4 × 1.50 should be viewed as a 24 × 15 computation. Then, to complete the problem apply a "test of reasonableness" (T of R). That is, ask yourself, "Where should I insert a decimal point within the intermediary figure to produce an answer that makes sense?" In this case, it would make sense to transform the intermediary figure of 360 (24 × 15) into 3.6. Here are some more examples:

Example #1

1.2 × 1.2

Step 1. Disregard the decimal points and think, "12 × 12."

Step 2. Multiply: 12 × 12 = 144 (intermediary product).

Step 3. Apply T of R: 144 is obviously too large to be the answer to 1.2 × 1.2. By "eyeballing" the problem, we know that the answer must be somewhere between 1 and 2.

Step 4. Insert a decimal point within the intermediary product, producing the answer 1.44.

Thought Process Summary

$$\begin{array}{r} 1.2 \\ \times 1.2 \\ \hline \end{array} \rightarrow \begin{array}{r} 12 \\ \times 12 \\ \hline 144 \end{array} \rightarrow 1.44$$

Example #2

48 ÷ 2.4

Step 1. Disregard the decimal point and think, "48 ÷ 24."

Step 2. Divide: 48 ÷ 24 = 2 (intermediary quotient).

Step 3. Apply T of R: 2 is obviously too small to be the answer to 48 ÷ 2.4. Affix one zero to the intermediary quotient, producing the answer 20.

Thought Process Summary

48 ÷ 2.4 → 48 ÷ 24 = 2 → 20

Example #3

930 ÷ 3.1

Step 1. Disregard the zero and decimal point, and think, "93 ÷ 31."

Step 2. Divide: 93 ÷ 31 = 3 (intermediary quotient).

Step 3. Apply T of R: 3 is obviously too small to be the answer to 930 ÷ 3.1; a quick estimate puts the answer around 300.

Step 4. Affix two zeroes to the intermediary quotient, producing the answer 300.

Thought Process Summary

$$930 \div 3.1 \rightarrow 93 \div 31 = 3 \rightarrow 300$$

Number-Power Note: In example #3 above, the "quick estimate" could be obtained by rounding the dividend to 900 and the divisor to 3.

Practice Exercises

Some of these exercises combine the two strategies learned to this point. Wherever possible, we will continue to build on tricks learned earlier—so be sure to do your best to remember them!

1. $80 \times 0.3 =$
2. $4.6 \times 200 =$
3. $700 \times 0.5 =$
4. $2.5 \times 300 =$
5. $3.9 \times 20 =$
6. $1.2 \times 120 =$
7. $1,800 \times 0.03 =$
8. $0.31 \times 30 =$
9. $720 \div 1.2 =$
10. $960 \div 3.2 =$
11. $150 \div 0.5 =$
12. $5,600 \div 1.4 =$
13. $81 \div 0.9 =$
14. $510 \div 1.7 =$

(See solutions on page 199)

Trick 3: Rapidly Multiply by 4 (or 0.4, 40, 400, etc.)

Strategy: Today's tricks are very easy to apply and very important. To multiply a number by 4, **double the number,** and then **double once again.** Remember to disregard any decimal points or zeroes when starting the calculation and to insert or affix to complete the calculation. From now on, our examples and practice exercises will be categorized as "Elementary" and "Brain Builders." In general, the elementary problems deal with the fundamentals, while the brain builders are more advanced and involve decimal points, larger numbers, and a higher degree of difficulty. Read on to see how Trick 3 is performed.

Elementary Example #1

32 × 4

Step 1. Double the 32: 32 × 2 = 64.

Step 2. Double the 64: 64 × 2 = 128 (the answer).

Thought Process Summary

$$\begin{array}{r} 32 \\ \times 4 \\ \hline \end{array} \rightarrow \begin{array}{r} 32 \\ \times 2 \\ \hline 64 \end{array} \rightarrow \begin{array}{r} 64 \\ \times 2 \\ \hline 128 \end{array}$$

Elementary Example #2

18 × 4

Step 1. Double the 18: 18 × 2 = 36.

Step 2. Double the 36: 36 × 2 = 72 (the answer).

Thought Process Summary

$$\begin{array}{r} 18 \\ \times 4 \\ \hline \end{array} \rightarrow \begin{array}{r} 18 \\ \times 2 \\ \hline 36 \end{array} \rightarrow \begin{array}{r} 36 \\ \times 2 \\ \hline 72 \end{array}$$

Brain Builder #1

2.4 × 40

Step 1. Disregard the decimal point and zero, and think, "24 × 4."

Step 2. Double the 24: 24 × 2 = 48.

Step 3. Double the 48: 48 × 2 = 96 (intermediary product).

Step 4. Apply T of R: A quick estimate puts the answer fairly close to 100. The intermediary product of 96 is also the answer.

Thought Process Summary

$$\begin{array}{r} 2.4 \\ \times 40 \\ \hline \end{array} \rightarrow \begin{array}{r} 24 \\ \times 4 \\ \hline \end{array} \rightarrow \begin{array}{r} 24 \\ \times 2 \\ \hline 48 \end{array} \rightarrow \begin{array}{r} 48 \\ \times 2 \\ \hline 96 \end{array}$$

Brain Builder #2

1,900 × 0.4

Step 1. Disregard the zeroes and decimal point, and think, "19 × 4."

Step 2. Double the 19: 19 × 2 = 38.

Step 3. Double the 38: 38 × 2 = 76 (intermediary product).

Step 4. Apply T of R: Since 0.4 is slightly below $\frac{1}{2}$, the answer must be somewhat below half of 1,900.

Step 5. Affix one zero to the intermediary product, producing the answer 760.

Thought Process Summary

$$\begin{array}{r} 1,900 \\ \times 0.4 \\ \hline \end{array} \rightarrow \begin{array}{r} 19 \\ \times 4 \\ \hline \end{array} \rightarrow \begin{array}{r} 19 \\ \times 2 \\ \hline 38 \end{array} \rightarrow \begin{array}{r} 38 \\ \times 2 \\ \hline 76 \end{array} \rightarrow 760$$

Number-Power Note: This trick is simple and should be obvious, yet surprisingly, many people do not use it. In fact, based on informal polls I've taken, only about one-third of my college students use this trick. One final question—how could you rapidly multiply by 8? That's right—simply double three times!

Elementary Exercises

From now on, the first 16 practice exercises will be labeled "Elementary," while the last 10 will be called "Brain Builders." I challenge you to try them all!

1. $35 \times 4 =$
2. $23 \times 4 =$
3. $14 \times 4 =$
4. $85 \times 4 =$
5. $4 \times 41 =$
6. $4 \times 26 =$
7. $4 \times 55 =$
8. $4 \times 72 =$
9. $61 \times 4 =$
10. $17 \times 4 =$
11. $95 \times 4 =$
12. $48 \times 4 =$
13. $4 \times 29 =$
14. $4 \times 83 =$
15. $4 \times 65 =$
16. $4 \times 53 =$

Brain Builders

1. $54 \times 40 =$
2. $360 \times 0.4 =$
3. $7.5 \times 4 =$
4. $0.15 \times 400 =$
5. $0.4 \times 910 =$
6. $40 \times 7.9 =$
7. $4 \times 570 =$
8. $400 \times 0.44 =$
9. $2.5 \times 40 =$
10. $98 \times 4 =$

(See solutions on page 199)

Parlor Trick #1: Finding the Fifth Root

What is the fifth root of a number? It's much like the square root, only taken a few steps further. For example, $7^5 = 7 \times 7 \times 7 \times 7 \times 7$, which equals 16,807. Therefore, the fifth root of 16,807 equals 7. Similarly, $24^5 = 24 \times 24 \times 24 \times 24 \times 24 = 7{,}962{,}624$. Therefore, the fifth root of 7,962,624 equals 24.

Ask someone to multiply a whole number (from 1 to 99) by itself five times, as shown above. An 8-digit calculator will be able to accommodate up to 39^5, whereas a 10-digit calculator will be able to accommodate up to 99^5. Since most calculators accommodate only 8 digits, you may wish to purchase one with a 10-digit display to allow up to 99^5. Obviously, don't watch as your volunteer is performing the calculation, but write down the final product when the calculation has been completed. It should take you no longer than a few seconds at that point to extract the fifth root.

Strategy: Let's take a couple of examples. Suppose someone has performed the requisite calculation and obtains a product of 32,768. Within about three seconds you will know that the fifth root of that number is 8.

Here's the trick—first of all, the ones-place digit in the product will automatically become the ones-place digit of the answer. Since 8 is the ones digit of 32,768, the ones digit of the fifth root will also be 8. Next, completely ignore the next four digits to the left of the ones digit (that is, the tens, hundreds, thousands, and ten-thousands digit). It may be easier to write the product down and cross out digits with a pencil than to just ignore them because you will then concentrate solely on the digits that remain. In the above case, no digits remain after crossing out the four digits, so the answer is simply 8.

To extract a two-digit fifth root, you'll need to have the following information memorized:

If no number remains, then the answer is a one-digit number.

If the remaining number is in the 1–30 range, the tens digit is 1.

If the remaining number is in the 30–230 range, the tens digit is 2.

If the remaining number is in the 230–1,000 range, the tens digit is 3.

If the remaining number is in the 1,000–3,000 range, the tens digit is 4.

If the remaining number is in the 3,000–7,500 range, the tens digit is 5.

If the remaining number is in the 7,500–16,000 range, the tens digit is 6.

If the remaining number is in the 16,000–32,000 range, the tens digit is 7.

If the remaining number is in the 32,000–57,000 range, the tens digit is 8.

If the remaining number is in the 57,000–99,000 range, the tens digit is 9.

Now let's take a number with a two-digit fifth root. Suppose someone performs the necessary calculation and arrives at 69,343,957. What will be the ones digit of the fifth root? That's right, it will be 7. Next, ignore or cross out the next four digits (that is, the 4395). What remains is the number 693.

This number is in the 230–1,000 range; therefore, the tens digit is 3, and the answer is 37. You might be wondering what to do if the remaining number is, for example, 230. Will the tens digit be 2 or 3? You don't have to worry because the remaining number will never be any of the border numbers.

When using the ranges above, it might be easiest to count off each number with your fingers, as follows: 1–30–230–1,000–3,000–7,500–16,000–32,000–57,000, until you reach the range containing the remaining number at hand. For example, we would have counted 1–30–230 in the above example, indicating a tens digit of 3. (We stop at 230 because the next number in the series, 1,000, exceeds the remaining number, 693.)

Try another exercise: Extract the fifth root of 7,339,040,224. You know that the ones digit of the answer is 4. Crossing out the next four digits, the remaining number is 73,390. Referring to the above ranges, you can see that the tens digit is 9, and the answer is 94.

Now it's your turn to extract the fifth root of:

A. 7,776
B. 844,596,301
C. 3,276,800,000
D. 459,165,024
E. 20,511,149
F. 371,293
G. 7,737,809,375
H. 2,887,174,368
I. 130,691,232
J. 79,235,168
K. 16,807
L. 9,509,900,499

(See solutions on page 224)

Trick 4: Rapidly Divide by 4 (or 0.4, 40, 400, etc.)

Strategy: This trick, too, is simple and should be obvious, yet many people do not use it. To divide a number by 4, **halve the number,** and then **halve once again.** Let's look at a few examples.

Elementary Example #1

84 ÷ 4

Step 1. Halve the 84: 84 ÷ 2 = 42

Step 2. Halve the 42: 42 ÷ 2 = 21 (the answer).

Thought Process Summary

84 ÷ 4 → 84 ÷ 2 = 42 → 42 ÷ 2 = 21

Elementary Example #2

76 ÷ 4

Step 1. Halve the 76: 76 ÷ 2 = 38

Step 2. Halve the 38: 38 ÷ 2 = 19 (the answer).

Thought Process Summary

76 ÷ 4 → 76 ÷ 2 = 38 → 38 ÷ 2 = 19

Brain Builder #1

620 ÷ 40

Step 1. Disregard the zeroes and think, "62 ÷ 4."

Step 2. Halve the 62: 62 ÷ 2 = 31

Step 3. Halve the 31: 31 ÷ 2 = 15.5 (intermediary quotient).

Step 4. Apply T of R: When dividing, you may cancel an equal number of zeroes from the dividend and divisor. Thus, 620 ÷ 40 = 62 ÷ 4. A quick estimate indicates that the intermediary quotient of 15.5 is the answer.

Thought Process Summary

620 ÷ 40 → 62 ÷ 2 = 31 → 31 ÷ 2 = 15.5

Brain Builder #2

9.2 ÷ 4

Step 1. Disregard the decimal point, converting the problem to 92 ÷ 4

Step 2. Halve the 92: 92 ÷ 2 = 46

Step 3. Halve the 46: 46 ÷ 2 = 23 (intermediary quotient).

Step 4. Apply T of R: A quick estimate puts the answer somewhere between 2 and 3

Step 5. Insert a decimal point within the intermediary quotient of 23, producing the answer 2.3

Thought Process Summary

$$\frac{9.2}{4} \rightarrow \frac{92}{4} \rightarrow \frac{92}{2} = 46 \rightarrow \frac{46}{2} = 23 \rightarrow 2.3$$

Number-Power Note: In certain instances, it really isn't necessary to disregard zeroes when starting a calculation. For example, 140 can be divided by 4 very quickly simply by just thinking, "140, 70, 35." Final note—how would you rapidly divide a number by 8? That's right, simply halve the number three times!

Elementary Exercises

Finish Day Two by "halving" yourself a ball with these:

1. $48 \div 4 =$
2. $68 \div 4 =$
3. $180 \div 4 =$
4. $132 \div 4 =$
5. $260 \div 4 =$
6. $96 \div 4 =$
7. $56 \div 4 =$
8. $88 \div 4 =$
9. $140 \div 4 =$
10. $220 \div 4 =$
11. $64 \div 4 =$
12. $72 \div 4 =$
13. $380 \div 4 =$
14. $340 \div 4 =$
15. $420 \div 4 =$
16. $52 \div 4 =$

Brain Builders

1. $440 \div 40 =$
2. $3,600 \div 400 =$
3. $112 \div 4 =$
4. $94 \div 4 =$
5. $540 \div 40 =$
6. $17.6 \div 0.4 =$
7. $14.4 \div 4 =$
8. $232 \div 4 =$
9. $81 \div 4 =$
10. $980 \div 400 =$

(See solutions on page 200)

Trick 5: Rapidly Multiply by 5 (or 0.5, 50, 500, etc.)

Strategy: This is our first trick using reciprocals. They can be very helpful in shortening a wide variety of problems—and they really do work! Reciprocals, as the term is used in this book, are two numbers that, when multiplied together, equal 10, 100, or any other power of 10. The numbers 5 and 2 are examples of reciprocals because they equal 10 when multiplied together. To multiply a number by 5, **divide the number by 2** and affix or insert any necessary zeroes or decimal point. (Be sure to disregard any decimal points or zeroes upon starting the calculation.) The assumption here is that it is easier to divide by 2 than to multiply by 5. Confused? You won't be after looking at the following examples.

Elementary Example #1

24 × 5

Step 1. Divide: 24 ÷ 2 = 12 (intermediary quotient).

Step 2. Apply T of R: 12 is obviously too small to be the answer to 24 × 5. A quick estimate puts the answer somewhere in the 100s.

Step 3. Affix one zero to the intermediary quotient, producing the answer 120.

Thought Process Summary

$$\begin{array}{r} 24 \\ \times 5 \\ \hline \end{array} \rightarrow 24 \div 2 = 12 \rightarrow 120$$

Elementary Example #2

46 × 5

Step 1. Divide: 46 ÷ 2 = 23 (intermediary quotient).

Step 2. Apply T of R: 23 is obviously too small to be the answer to 46 × 5. A quick estimate puts the answer somewhere in the 200s.

Step 3. Affix one zero to the intermediary quotient, producing the answer 230.

Thought Process Summary

$$\begin{array}{r} 46 \\ \times 5 \\ \hline \end{array} \rightarrow 46 \div 2 = 23 \rightarrow 230$$

Brain Builder #1

35 × 50

Step 1. Disregard the zero and think, "35 × 5."

Step 2. Divide: 35 ÷ 2 = 17.5 (intermediary quotient).

Step 3. Apply T of R: 17.5 is obviously too small to be the answer to 35 × 50. A quick estimate puts the answer between 1,500 and 2,000.

Step 4. Eliminate the decimal point from the intermediary quotient and affix one zero, producing the answer 1,750.

Thought Process Summary

$$\begin{array}{r} 35 \\ \times 50 \\ \hline \end{array} \rightarrow \begin{array}{r} 35 \\ \times 5 \\ \hline \end{array} \rightarrow 35 \div 2 = 17.5 \rightarrow 1,750$$

Brain Builder #2

36.2 × 5

Step 1. Disregard the decimal point, and think, "362 × 5."

Step 2. Divide: 362 ÷ 2 = 181 (intermediary quotient).

Step 3. Apply T of R: 181 seems reasonable as the answer to 36.2 × 5, and in fact is the answer.

Thought Process Summary

$$\begin{array}{r} 36.2 \\ \times 5 \\ \hline \end{array} \rightarrow \begin{array}{r} 362 \\ \times 5 \\ \hline \end{array} \rightarrow 362 \div 2 = 181$$

Number-Power Note: This trick can also be used to multiply a number by $\frac{1}{2}$, since $\frac{1}{2}$ is the same as 0.5. However, for purposes of simplicity, any exercises in this book that involve other than whole numbers will be presented in decimal rather than fraction form.

Elementary Exercises

You won't believe how quickly you'll be able to do these exercises.

1. $16 \times 5 =$
2. $38 \times 5 =$
3. $88 \times 5 =$
4. $42 \times 5 =$
5. $5 \times 74 =$
6. $5 \times 58 =$
7. $5 \times 22 =$
8. $5 \times 76 =$
9. $62 \times 5 =$
10. $28 \times 5 =$
11. $66 \times 5 =$
12. $94 \times 5 =$
13. $5 \times 54 =$
14. $5 \times 82 =$
15. $5 \times 96 =$
16. $5 \times 44 =$

Brain Builders

1. $85 \times 5 =$
2. $49 \times 5 =$
3. $33 \times 5 =$
4. $97 \times 5 =$
5. $50 \times 55 =$
6. $0.5 \times 12.2 =$
7. $500 \times 0.79 =$
8. $5 \times 510 =$
9. $2.1 \times 50 =$
10. $6.8 \times 500 =$

(See solutions on page 200)

Trick 6: Rapidly Divide by 5 (or 0.5, 50, 500, etc.)

Strategy: Reciprocals work as effectively with division as with multiplication. To divide a number by 5, **multiply the number by 2** and affix or insert any necessary zeroes or decimal point. This technique follows the same general procedure that you learned in Trick 5 and that you will continue to apply with other multiplication and division techniques. That is, you first **operate** (multiply or divide); then you **think** (apply a test of reasonableness), and finally you **adjust,** if necessary (insert or affix a decimal point or zeroes). You'll be amazed to see how easily the following examples are operated upon.

Elementary Example #1

38 ÷ 5

Step 1. Multiply: 38 × 2 = 76 (intermediary product).

Step 2. Apply T of R: 76 is obviously too large to be the answer to 38 ÷ 5. A quick estimate puts the answer somewhere between 7 and 8.

Step 3. Insert a decimal point within the intermediary product, producing the answer 7.6.

Thought Process Summary

$$38 \div 5 \rightarrow \begin{array}{r} 38 \\ \times 2 \\ \hline 76 \end{array} \rightarrow 7.6$$

Elementary Example #2

85 ÷ 5

Step 1. Multiply: 85 × 2 = 170 (intermediary product).

Step 2. Apply T of R: 170 is obviously too large to be the answer to 85 ÷ 5. A quick estimate puts the answer somewhere between 10 and 20.

Step 3. Insert a decimal point within the intermediary product, producing the answer 17.0 (or simply 17).

Thought Process Summary

$$85 \div 5 \rightarrow \begin{array}{r} 85 \\ \times 2 \\ \hline 170 \end{array} \rightarrow 17$$

Brain Builder #1

245 ÷ 50

Step 1. Multiply: 245 × 2 = 490 (intermediary product).

Step 2. Apply T of R: 490 is obviously too large to be the answer to 245 ÷ 50. A quick estimate puts the answer just below 5.

Step 3. Insert a decimal point within the intermediary product, producing the answer 4.90 (or simply 4.9).

Thought Process Summary

$245 \div 50 \rightarrow 245 \times 2 = 490 \rightarrow 4.9$

Brain Builder #2

44.4 ÷ 5

Step 1. Disregard the decimal point, converting the problem to 444 ÷ 5.

Step 2. Multiply: 444 × 2 = 888 (intermediary product).

Step 3. Apply T of R: 888 is obviously too large to be the answer to 44.4 ÷ 5. A quick estimate puts the answer just below 9.

Step 4. Insert a decimal point within the intermediary product, producing the answer 8.88.

Thought Process Summary

$44.4 \div 5 \rightarrow 444 \div 5 \rightarrow 444 \times 2 = 888 \rightarrow 8.88$

Number-Power Note: When applying a test of reasonableness to division problems, try to use the inverse of division—multiplication. For example, in Brain Builder #1 above, ask yourself, "50 times what equals 245?" With a moment's thought, you can place the answer just below 5.

Elementary Exercises

How should you approach the following exercises? Divide and conquer, of course!

1. 27 ÷ 5 =
2. 53 ÷ 5 =
3. 72 ÷ 5 =
4. 67 ÷ 5 =
5. 118 ÷ 5 =
6. 95 ÷ 5 =
7. 41 ÷ 5 =
8. 49 ÷ 5 =
9. 122 ÷ 5 =
10. 14 ÷ 5 =
11. 76 ÷ 5 =
12. 81 ÷ 5 =

13. $33 \div 5 =$
14. $58 \div 5 =$
15. $98 \div 5 =$
16. $64 \div 5 =$

Brain Builders

1. $230 \div 50 =$
2. $18.5 \div 5 =$
3. $8,300 \div 500 =$
4. $33.3 \div 5 =$
5. $190 \div 50 =$
6. $4.2 \div 5 =$
7. $43.5 \div 0.5 =$
8. $920 \div 50 =$
9. $610 \div 500 =$
10. $4.6 \div 5 =$

(See solutions on page 200)

Trick 7: Rapidly Square Any Number Ending in 5

Strategy: This trick is one of the oldest in the book, and one of the best! To square a number that ends in 5, first **multiply the tens digit by the next whole number**. To that product, **affix the number 25**. The number to affix (25) is easy to remember, because $5^2 = 25$. Although a calculation such as 7.5×750 is technically not a square, it too can be solved using this technique. This trick will also work for numbers with more than two digits. Read on to see how this marvelous trick works.

Elementary Example #1

15^2

Step 1. Multiply: $1 \times 2 = 2$.

Step 2. Affix 25: 225 (the answer).

Thought Process Summary

$$\begin{array}{r} 15 \\ \times 15 \\ \hline \end{array} \rightarrow \begin{array}{r} 1 \\ \times 2 \\ \hline 2 \end{array} \rightarrow 225$$

Elementary Example #2

65^2

Step 1. Multiply: $6 \times 7 = 42$.

Step 2. Affix 25: 4,225 (the answer).

Thought Process Summary

$$\begin{array}{r} 65 \\ \times 65 \\ \hline \end{array} \rightarrow \begin{array}{r} 6 \\ \times 7 \\ \hline 42 \end{array} \rightarrow 4,225$$

Brain Builder #1

450^2

Step 1. Disregard the zero and think, "45 squared."

Step 2. Multiply: 4 × 5 = 20.

Step 3. Affix 25: 2,025 (intermediary product).

Step 4. Apply T of R: For each zero initially disregarded in a squaring problem, two must eventually be affixed to obtain the product.

Step 5. Affix two zeroes to the intermediary product, producing the answer 202,500.

Thought Process Summary

$$\begin{array}{r} 450 \\ \times 450 \\ \hline \end{array} \rightarrow \begin{array}{r} 45 \\ \times 45 \\ \hline \end{array} \rightarrow \begin{array}{r} 4 \\ \times 5 \\ \hline 20 \end{array} \rightarrow 2,025 \rightarrow 202,500$$

Brain Builder #2

7.5 × 750

Step 1. Disregard the decimal point and zero, and think, "75 squared."

Step 2. Multiply: 7 × 8 = 56.

Step 3. Affix 25: 5,625 (intermediary product).

Step 4. Apply T of R: A quick estimate puts the answer in the 5,000s. The intermediary product of 5,625 is therefore the answer.

Thought Process Summary

$$\begin{array}{r} 7.5 \\ \times 750 \\ \hline \end{array} \rightarrow \begin{array}{r} 75 \\ \times 75 \\ \hline \end{array} \rightarrow \begin{array}{r} 7 \\ \times 8 \\ \hline 56 \end{array} \rightarrow 5,625$$

Brain Builder #3

115^2

Step 1. Multiply: $11 \times 12 = 132$.

Step 2. Affix 25: 13,225 (the answer).

Thought Process Summary

$$\begin{array}{r} 115 \\ \times 115 \\ \hline \end{array} \rightarrow \begin{array}{r} 11 \\ \times 12 \\ \hline 132 \end{array} \rightarrow 13,225$$

Number-Power Notes: Trick 23 involves rapidly squaring any two-digit number **beginning** in 5, and Trick 20 is a variation on Trick 7.

Elementary Exercises

Don't be confused by the two ways these squaring exercises are presented.

1. $35 \times 35 =$
2. $85 \times 85 =$
3. $95 \times 95 =$
4. $25 \times 25 =$
5. $55 \times 55 =$
6. $75 \times 75 =$
7. $45 \times 45 =$
8. $15 \times 15 =$
9. $65^2 =$
10. $95^2 =$
11. $85^2 =$
12. $35^2 =$
13. $25^2 =$
14. $55^2 =$
15. $75^2 =$
16. $45^2 =$

Brain Builders

1. $105 \times 105 =$
2. $3.5 \times 350 =$
3. $750^2 =$
4. $0.85 \times 85 =$
5. $65 \times 6.5 =$
6. $150 \times 15 =$
7. $11.5^2 =$
8. $5.5 \times 550 =$
9. $0.45 \times 0.45 =$
10. $950 \times 9.5 =$

(See solutions on page 201)

Mathematical Curiosity #1

$12{,}345{,}679 \times 9 = 111{,}111{,}111$
$12{,}345{,}679 \times 18 = 222{,}222{,}222$
$12{,}345{,}679 \times 27 = 333{,}333{,}333$
$12{,}345{,}679 \times 36 = 444{,}444{,}444$
$12{,}345{,}679 \times 45 = 555{,}555{,}555$
$12{,}345{,}679 \times 54 = 666{,}666{,}666$
$12{,}345{,}679 \times 63 = 777{,}777{,}777$
$12{,}345{,}679 \times 72 = 888{,}888{,}888$
$12{,}345{,}679 \times 81 = 999{,}999{,}999$
$12{,}345{,}679 \times 999{,}999{,}999 = 12{,}345{,}678{,}987{,}654{,}321$

Trick 8: Rapidly Multiply Any Two-Digit Number by 11 (or 0.11, 1.1, 110, etc.)

Strategy: The "11 trick" is the most popular trick of them all—and one of the most useful. To multiply a two-digit number by 11, first **write the number**, leaving some space between the two digits. Then **insert the sum of the number's two digits** in between the two digits themselves. You will have to carry when the sum of the digits exceeds 9. This trick is almost too good to be true, as you'll see in the following examples.

Elementary Example #1

24 × 11

Step 1. Write the multiplicand in the answer space, leaving some room between the two digits: 2 ? 4 (intermediary product).

Step 2. Add the two digits of the multiplicand: 2 + 4 = 6.

Step 3. Insert the 6 within the intermediary product, producing the answer 264.

Thought Process Summary

24 × 11 → 2 ? 4 → 2 + 4 = 6 → 264

Elementary Example #2

76 × 11

Step 1. Write the multiplicand in the answer space, leaving some room between the two digits: 7 ? 6 (intermediary product).

Step 2. Add the two digits of the multiplicand: 7 + 6 = 13.

Step 3. Insert the ones digit of the 13 within the intermediary product, producing a second intermediary product of 736.

Step 4. Because the two digits of the multiplicand total more than 9, one must "carry," converting 736 to the answer, 836.

Thought Process Summary

$$\begin{array}{r} 76 \\ \times 11 \\ \hline \end{array} \rightarrow 7\,\underline{?}\,6 \rightarrow \begin{array}{r} 7 \\ +6 \\ \hline 13 \end{array} \rightarrow 736 \rightarrow 836$$

Brain Builder #1

53 × 110

Step 1. Disregard the zero, and think, "53 × 11."

Step 2. Follow steps 1, 2, and 3 of the preceding Elementary Examples, producing an intermediary product of 583.

Step 3. Apply T of R: The zero that was disregarded must be reaffixed, producing the answer 5,830.

Thought Process Summary

$$\begin{array}{r} 53 \\ \times 110 \\ \hline \end{array} \rightarrow \begin{array}{r} 53 \\ \times 11 \\ \hline \end{array} \rightarrow 5\,\underline{?}\,3 \rightarrow \begin{array}{r} 5 \\ +3 \\ \hline 8 \end{array} \rightarrow 583 \rightarrow 5,830$$

Brain Builder #2

8.7 × 1.1

Step 1. Disregard the decimal points and think, "87 × 11."

Step 2. Follow steps 1, 2, and 3 of the preceding Elementary Examples, producing an intermediary product of 957.

Step 3. Apply T of R: A quick estimate puts the answer at about 9 or 10.

Step 4. Insert a decimal point within the intermediary product, producing the answer 9.57.

Thought Process Summary

$$\begin{array}{r} 8.7 \\ \times 1.1 \\ \hline \end{array} \rightarrow \begin{array}{r} 87 \\ \times 11 \\ \hline \end{array} \rightarrow 8\,\underline{?}\,7 \rightarrow \begin{array}{r} 8 \\ +7 \\ \hline 15 \end{array} \rightarrow 857 \rightarrow 957 \rightarrow 9.57$$

Number-Power Note: Trick 59 involves rapidly multiplying a three-digit or larger number by 11. An alternative but less efficient method to multiply by 11 is to multiply the number by 10 and then add the number itself. For example, in Elementary Example #1 above, $24 \times 11 = (24 \times 10) + 24 = 264$.

Elementary Exercises

You should be able to float through these exercises with the greatest of ease.

1. $62 \times 11 =$
2. $18 \times 11 =$
3. $35 \times 11 =$
4. $81 \times 11 =$
5. $11 \times 26 =$
6. $11 \times 44 =$
7. $11 \times 58 =$
8. $11 \times 92 =$
9. $17 \times 11 =$
10. $69 \times 11 =$
11. $31 \times 11 =$
12. $74 \times 11 =$
13. $11 \times 96 =$
14. $11 \times 39 =$
15. $11 \times 47 =$
16. $11 \times 99 =$

Brain Builders

1. $2.7 \times 110 =$
2. $65 \times 1.1 =$
3. $8.3 \times 1.1 =$
4. $0.56 \times 1{,}100 =$
5. $1.1 \times 130 =$
6. $110 \times 330 =$
7. $110 \times 7.8 =$
8. $1.1 \times 2.2 =$
9. $940 \times 11 =$
10. $4.9 \times 110 =$

(See solutions on page 201)

Trick 9: Rapidly Multiply by 25 (or 0.25, 2.5, 250, etc.)

Strategy: Let's begin day five with another reciprocal technique that will come in handy time and time again. To multiply a number by 25, **divide the number by 4** and affix or insert any necessary zeroes or decimal point. Here are some examples that illustrate this very practical trick.

Elementary Example #1

28 × 25

Step 1. Divide: 28 ÷ 4 = 7 (intermediary quotient).

Step 2. Apply T of R: 7 and 70 both seem too small to be the answer to 28 × 25.

Step 3. Affix two zeroes to the intermediary quotient, producing the answer 700.

Thought Process Summary

$$\begin{array}{r} 28 \\ \times 25 \\ \hline \end{array} \rightarrow 28 \div 4 = 7 \rightarrow 700$$

Elementary Example #2

76 × 25

Step 1. Divide: 76 ÷ 4 = 19 (intermediary quotient).

Step 2. Apply T of R: 19 and 190 both seem too small to be the answer to 76 × 25.

Step 3. Affix two zeroes to the intermediary quotient, producing the answer 1,900.

Thought Process Summary

$$\begin{array}{r} 76 \\ \times 25 \\ \hline \end{array} \rightarrow 76 \div 4 = 19 \rightarrow 1{,}900$$

Brain Builder #1

36 × 250

Step 1. Divide: 36 ÷ 4 = 9 (intermediary quotient).

Step 2. Apply T of R: 9, 90, and 900 all seem too small to be the answer to 36 × 250.

Step 3. Affix three zeroes to the intermediary quotient, producing the answer 9,000.

Thought Process Summary

$$\begin{array}{r} 36 \\ \times 250 \\ \hline \end{array} \rightarrow 36 \div 4 = 9 \rightarrow 9{,}000$$

Brain Builder #2

420 × 2.5

Step 1. Disregard the zero and decimal point, and think, "42 × 25."

Step 2. Divide: 42 ÷ 4 = 10.5 (intermediary quotient).

Step 3. Apply T of R: A quick estimate puts the answer around 1,000.

Step 4. Eliminate the decimal point from the intermediary quotient and affix one zero to produce the answer 1,050.

Thought Process Summary

$$\begin{array}{r} 420 \\ \times 2.5 \\ \hline \end{array} \rightarrow \begin{array}{r} 42 \\ \times 25 \\ \hline \end{array} \rightarrow 42 \div 4 = 10.5 \rightarrow 1{,}050$$

Number-Power Note: Remember that Trick 4 recommends that you divide by 4 by halving, and then halving again.

Elementary Exercises

You can combine Tricks 4 and 9 when working these exercises.

1. $12 \times 25 =$
2. $44 \times 25 =$
3. $52 \times 25 =$
4. $16 \times 25 =$
5. $25 \times 64 =$
6. $25 \times 88 =$
7. $25 \times 24 =$
8. $25 \times 56 =$
9. $92 \times 25 =$
10. $68 \times 25 =$
11. $48 \times 25 =$
12. $80 \times 25 =$
13. $25 \times 32 =$
14. $25 \times 84 =$
15. $25 \times 72 =$
16. $25 \times 96 =$

Brain Builders

1. $34 \times 25 =$
2. $78 \times 25 =$
3. $58 \times 25 =$
4. $14 \times 25 =$
5. $25 \times 7.4 =$
6. $250 \times 28 =$
7. $2.5 \times 86 =$
8. $0.25 \times 600 =$
9. $25 \times 8.4 =$
10. $250 \times 18 =$

(See solutions on page 201)

Trick 10: Rapidly Divide by 25 (or 0.25, 2.5, 250, etc.)

Strategy: Perhaps you've already guessed this trick's shortcut. To divide a number by 25, **multiply the number by 4**, and affix or insert any necessary zeroes or decimal point. Dividing by 25 is even easier than multiplying by 25, as you'll see in the following examples.

Elementary Example #1
700 ÷ 25

Step 1. Disregard the zeroes and multiply: $7 \times 4 = 28$ (intermediary product).

Step 2. Apply T of R: 28 is not only in the ballpark, it is the answer. (Refer to the Number-Power Note for further explanation.)

Thought Process Summary

$700 \div 25 \rightarrow 7 \div 25 \rightarrow 7 \times 4 = 28$

Elementary Example #2

210 ÷ 25

Step 1. Disregard the zero and multiply: 21 × 4 = 84 (intermediary product).

Step 2. Apply T of R: A quick estimate puts the answer between 8 and 9.

Step 3. Insert a decimal point within the intermediary product, producing the answer 8.4.

Thought Process Summary

$$210 \div 25 \rightarrow 21 \div 25 \rightarrow \begin{array}{r} 21 \\ \times 4 \\ \hline 84 \end{array} \rightarrow 8.4$$

Brain Builder #1

450 ÷ 2.5

Step 1. Disregard the zero and decimal point, and multiply: 45 × 4 = 180 (intermediary product).

Step 2. Apply T of R: A quick estimate puts the answer somewhere between 100 and 200. Thus, the intermediary product (180) is the answer.

Thought Process Summary

$$450 \div 2.5 \rightarrow 45 \div 25 \rightarrow \begin{array}{r} 45 \\ \times 4 \\ \hline 180 \end{array}$$

Brain Builder #2

5,200 ÷ 250

Step 1. Disregard the zeroes and multiply: 52 × 4 = 208 (intermediary product).

Step 2. Apply T of R: Remove one zero each from the original dividend and divisor, and try to arrive at an estimate for 520 ÷ 25. It appears to be around 20.

Step 3. Insert a decimal point within the intermediary product, producing the answer 20.8.

Thought Process Summary

$$5{,}200 \div 250 \rightarrow 52 \div 25 \rightarrow \begin{array}{r} 52 \\ \times 4 \\ \hline 208 \end{array} \rightarrow 20.8$$

Number-Power Notes: Remember that Trick 3 recommends that you multiply by 4 by doubling, and then doubling again. Also, to obtain an estimate for $700 \div 25$ in Elementary Example #1, you could work backwards and think, "25 goes into 100 four times, and 100 goes into 700 seven times; 4 times 7 equals 28."

Elementary Exercises

You can combine Tricks 3 and 10 when working through these exercises.

1. $80 \div 25 =$
2. $300 \div 25 =$
3. $180 \div 25 =$
4. $2{,}400 \div 25 =$
5. $320 \div 25 =$
6. $650 \div 25 =$
7. $1{,}200 \div 25 =$
8. $850 \div 25 =$
9. $95 \div 25 =$
10. $400 \div 25 =$
11. $9{,}000 \div 25 =$
12. $1{,}500 \div 25 =$
13. $500 \div 25 =$
14. $2{,}200 \div 25 =$
15. $750 \div 25 =$
16. $270 \div 25 =$

Brain Builders

1. $350 \div 2.5 =$
2. $111 \div 25 =$
3. $2.3 \div 2.5 =$
4. $550 \div 25 =$
5. $1{,}700 \div 250 =$
6. $4.2 \div 0.25 =$
7. $222 \div 25 =$
8. $820 \div 2.5 =$
9. $130 \div 25 =$
10. $3{,}300 \div 250 =$

(See solutions on page 202)

Trick 11: Rapidly Multiply Any One- or Two-Digit Number by 99 (or 0.99, 9.9, 990, etc.)

Strategy: Nine is by far the most mysterious and magical number. Here is a trick involving two 9s. To multiply a one- or two-digit number by 99, first **subtract 1 from the number** to obtain the **left-hand portion** of the answer. Then, **subtract the number from 100** to obtain the **right-hand portion**. (Hint: You will learn from Trick 29 that it is faster to subtract by adding. For example, to subtract 88 from 100, ask yourself, "88 plus what equals 100?" to obtain the answer, 12.) When you've reviewed the examples below, you'll see that this trick is a lot easier to apply than it sounds.

Elementary Example #1

15 × 99

Step 1. Subtract: $15 - 1 = 14$ (left-hand portion of the answer).

Step 2. Subtract: $100 - 15 = 85$ (right-hand portion of the answer).

Step 3. Combine: 1,485 is the answer.

Thought Process Summary

$$\begin{array}{r} 15 \\ \times 99 \\ \hline \end{array} \rightarrow \begin{array}{r} 15 \\ -1 \\ \hline 14 \end{array} \rightarrow \begin{array}{r} 100 \\ -15 \\ \hline 85 \end{array} \rightarrow 1,485$$

Elementary Example #2

7 × 99

Step 1. Subtract: $7 - 1 = 6$ (left-hand portion of the answer).

Step 2. Subtract: $100 - 7 = 93$ (right-hand portion of the answer).

Step 3. Combine: 693 is the answer.

Thought Process Summary

$$\begin{array}{r} 7 \\ \times 99 \\ \hline \end{array} \rightarrow \begin{array}{r} 7 \\ -1 \\ \hline 6 \end{array} \rightarrow \begin{array}{r} 100 \\ -7 \\ \hline 93 \end{array} \rightarrow 693$$

Brain Builder #1

2.8 × 9.9

Step 1. Disregard decimal points and subtract: 28 − 1 = 27 (left-hand portion of the answer).

Step 2. Subtract: 100 − 28 = 72 (right-hand portion of the answer).

Step 3. Combine: 2,772 (intermediary product).

Step 4. Apply T of R: 9.9 is almost 10, so 2.8 × 9.9 must equal just under 28. Insert a decimal point within the intermediary product to obtain the answer, 27.72.

Thought Process Summary

$$\begin{array}{r} 2.8 \\ \times 9.9 \\ \hline \end{array} \rightarrow \begin{array}{r} 28 \\ -1 \\ \hline 27 \end{array} \rightarrow \begin{array}{r} 100 \\ -28 \\ \hline 72 \end{array} \rightarrow 2,772 \rightarrow 27.72$$

Brain Builder #2

430 × 0.99

Step 1. Disregard the zero and decimal point, and subtract: 43 − 1 = 42 (left-hand portion of the answer).

Step 2. Subtract: 100 − 43 = 57 (right-hand portion of the answer).

Step 3. Combine: 4,257 (intermediary product).

Step 4. Apply T of R: 0.99 is almost 1, so 430 × 0.99 must equal just under 430. Insert a decimal point within the intermediary product to obtain the answer, 425.7.

Thought Process Summary

$$\begin{array}{r} 430 \\ \times 0.99 \\ \hline \end{array} \rightarrow \begin{array}{r} 43 \\ -1 \\ \hline 42 \end{array} \rightarrow \begin{array}{r} 100 \\ -43 \\ \hline 57 \end{array} \rightarrow 4,257 \rightarrow 425.7$$

Number-Power Note: Trick 60 shows how to rapidly **divide** by 9, 99, and so forth.

Elementary Exercises

In doing these exercises, remember to obtain the left-hand portion of the answer before computing the right-hand portion.

1. $60 \times 99 =$
2. $75 \times 99 =$
3. $9 \times 99 =$
4. $88 \times 99 =$
5. $99 \times 35 =$
6. $99 \times 61 =$
7. $99 \times 66 =$
8. $99 \times 48 =$
9. $80 \times 99 =$
10. $22 \times 99 =$
11. $4 \times 99 =$
12. $54 \times 99 =$
13. $99 \times 83 =$
14. $99 \times 39 =$
15. $99 \times 97 =$
16. $99 \times 11 =$

Brain Builders

1. $5.2 \times 990 =$
2. $91 \times 9.9 =$
3. $0.77 \times 99 =$
4. $260 \times 0.99 =$
5. $9.9 \times 200 =$
6. $990 \times 330 =$
7. $99 \times 7.2 =$
8. $0.99 \times 440 =$
9. $0.57 \times 9.9 =$
10. $3 \times 990 =$

(See solutions on page 202)

Mathematical Curiosity #2

$$987,654,321 \times 9 = 8,888,888,889$$
$$987,654,321 \times 18 = 17,777,777,778$$
$$987,654,321 \times 27 = 26,666,666,667$$
$$987,654,321 \times 36 = 35,555,555,556$$
$$987,654,321 \times 45 = 44,444,444,445$$
$$987,654,321 \times 54 = 53,333,333,334$$
$$987,654,321 \times 63 = 62,222,222,223$$
$$987,654,321 \times 72 = 71,111,111,112$$
$$987,654,321 \times 81 = 80,000,000,001$$

(Also note: In each case, the first and last digit of the product is the same as the multiplier.)

Trick 12: Rapidly Multiply Any One- or Two-Digit Number by 101 (or 1.01, 10.1, 1,010, etc.)

Strategy: Day six ends with the easiest trick of them all. To multiply a one-digit number by 101, **write down the one-digit number twice**, and insert a zero in between. For example, $101 \times 7 = 707$. To multiply a two-digit number by 101, **write down the two-digit number twice**, and you have the answer! For example, $36 \times 101 = 3,636$. You probably don't need more examples to understand this trick, but here are some more, anyway.

Elementary Example #1

4 × 101

Step 1. Write down the 4 twice, and insert a zero: 404 (the answer).

Thought Process Summary

$$\begin{array}{r} 4 \\ \underline{\times 101} \end{array} \quad \rightarrow \quad 404$$

Elementary Example #2

27 × 101

Step 1. Write down the 27 twice: 2,727 (the answer).

Thought Process Summary

$$\begin{array}{r} 27 \\ \underline{\times 101} \end{array} \quad \rightarrow \quad 2,727$$

Brain Builder #1

56 × 1.01

Step 1. Write down the 56 twice: 5,656 (intermediary product).

Step 2. Apply T of R: A quick estimate puts the answer just above 56.

Step 3. Insert a decimal point within the intermediary product, producing the answer 56.56.

Thought Process Summary

$56 \times 1.01 \rightarrow 5,656 \rightarrow 56.56$

Brain Builder #2

740 × 10.1

Step 1. Disregard the zero, and write down the 74 twice: 7,474 (intermediary product).

Step 2. Apply T of R: A quick estimate puts the answer just above 7,400. Accordingly, 7,474 appears to be the answer.

Thought Process Summary

$740 \times 10.1 \rightarrow 7,474$

Number-Power Note: To multiply a two-digit number by 1,001, perform the same steps as for 101, but insert a zero in the middle. For example, $49 \times 1,001 = 49,049$. However, to multiply a three-digit number by 1,001, simply write down the three-digit number twice! For example, $417 \times 1,001 = 417,417$.

Elementary Exercises

You'll be able to work through these exercises as quickly as you can write the answers!

1. $15 \times 101 =$
2. $62 \times 101 =$
3. $39 \times 101 =$
4. $8 \times 101 =$
5. $101 \times 93 =$
6. $101 \times 41 =$
7. $101 \times 87 =$
8. $101 \times 70 =$
9. $12 \times 101 =$
10. $45 \times 101 =$
11. $81 \times 101 =$
12. $23 \times 101 =$
13. $101 \times 6 =$
14. $101 \times 78 =$
15. $101 \times 32 =$
16. $101 \times 99 =$

Brain Builders

1. $4.8 \times 1.01 =$
2. $630 \times 0.101 =$
3. $0.36 \times 1,010 =$
4. $8.5 \times 10.1 =$
5. $101 \times 110 =$
6. $1.01 \times 920 =$
7. $10.1 \times 30 =$
8. $1,010 \times 1.9 =$
9. $890 \times 1.01 =$
10. $5.1 \times 101 =$

(See solutions on page 202)

Trick 13: Rapidly Multiply Two Numbers Whose Difference Is 2

Strategy: Before learning this trick, you should review the squares table on page 11. To multiply two numbers whose difference is 2, first **square the number between the two**. Then, **subtract 1** from the product to obtain the answer. Let's look at some examples of this trick.

Elementary Example #1

11 × 13

Step 1. Square the number in the middle: $12^2 = 144$ (intermediary product).

Step 2. Subtract 1 from the intermediary product: $144 - 1 = 143$, which is the answer. (Note: Why not use Trick 8—the "11 trick"—to check your answer?)

Thought Process Summary

$$\begin{array}{r} 11 \\ \times 13 \\ \hline \end{array} \rightarrow \begin{array}{r} 12 \\ \times 12 \\ \hline 144 \end{array} \rightarrow \begin{array}{r} 144 \\ -1 \\ \hline 143 \end{array}$$

Elementary Example #2

24 × 26

Step 1. Square the number in the middle: $25^2 = 625$ (intermediary product). (Did you remember to use Trick 7 to square the 25?)

Step 2. Subtract 1 from the intermediary product: $625 - 1 = 624$, which is the answer.

Thought Process Summary

$$\begin{array}{r} 24 \\ \times 26 \\ \hline \end{array} \rightarrow \begin{array}{r} 25 \\ \times 25 \\ \hline 625 \end{array} \rightarrow \begin{array}{r} 625 \\ -1 \\ \hline 624 \end{array}$$

Brain Builder #1

1.9 × 210

Step 1. Disregard the decimal point and zero, and think, "19 × 21."

Step 2. Square the number in the middle: $20^2 = 400$ (intermediary product).

Step 3. Subtract 1 from the intermediary product: $400 - 1 = 399$ (revised intermediary product).

Step 4. Apply T of R: A quick estimate puts the answer around 400, so the revised intermediary product of 399 is the answer.

Thought Process Summary

$$\begin{array}{r} 1.9 \\ \times 210 \\ \hline \end{array} \rightarrow \begin{array}{r} 19 \\ \times 21 \\ \hline \end{array} \rightarrow \begin{array}{r} 20 \\ \times 20 \\ \hline 400 \end{array} \rightarrow \begin{array}{r} 400 \\ -1 \\ \hline 399 \end{array}$$

Brain Builder #2

1.4 × 1.2

Step 1. Disregard the decimal points and think, "14 × 12."

Step 2. Square the number in the middle: $13^2 = 169$ (intermediary product).

Step 3. Subtract 1 from the intermediary product: $169 - 1 = 168$ (revised intermediary product).

Step 4. Apply T of R: A quick estimate puts the answer somewhere between 1 and 2.

Step 5. Insert a decimal point within the revised intermediary product, producing the answer 1.68.

Thought Process Summary

$$\begin{array}{r} 1.4 \\ \times 1.2 \\ \hline \end{array} \rightarrow \begin{array}{r} 14 \\ \times 12 \\ \hline \end{array} \rightarrow \begin{array}{r} 13 \\ \times 13 \\ \hline 169 \end{array} \rightarrow \begin{array}{r} 169 \\ -1 \\ \hline 168 \end{array} \rightarrow 1.68$$

Number-Power Note: Trick 26 is a variation on this technique.

Elementary Exercises

The better you are at squaring, the easier you'll find these exercises.

1. $15 \times 13 =$
2. $17 \times 15 =$
3. $29 \times 31 =$
4. $14 \times 16 =$
5. $36 \times 34 =$
6. $79 \times 81 =$
7. $99 \times 101 =$
8. $54 \times 56 =$
9. $13 \times 11 =$
10. $12 \times 14 =$
11. $21 \times 19 =$
12. $16 \times 18 =$
13. $59 \times 61 =$
14. $26 \times 24 =$
15. $19 \times 17 =$
16. $84 \times 86 =$

Brain Builders

1. $2.1 \times 19 =$
2. $74 \times 7.6 =$
3. $110 \times 1.3 =$
4. $1.7 \times 1.5 =$
5. $0.44 \times 460 =$
6. $9.9 \times 101 =$
7. $140 \times 120 =$
8. $49 \times 0.51 =$
9. $1.9 \times 17 =$
10. $640 \times 6.6 =$

(See solutions on page 203)

Parlor Trick #2: The Perpetual Calendar

This trick will enable you to determine, within about 15 or 30 seconds, the day of the week, given any date in the twentieth century.

At first, the amount of information to memorize may seem rather overwhelming. However, with some patience, practice, and concentration, you will be able to master this excellent parlor trick. You may even find it useful on a day-to-day basis!

Strategy: You will need to memorize the perpetual calendar formula:

$$\frac{\text{Year} + \text{Year}/4 + \text{Date of month} + \text{Factor for month}}{7}$$

What matters is not the quotient of this division, but the remainder, as will be explained.

The Year in the equation above consists of the last two digits. For example, the last two digits of 1935 are 35. When calculating Year/4, divide 4 into the last two digits of the year. It's important that you remember to **round down** to the nearest whole number. For example, if the year under consideration is 1935, take 35 and divide it by 4, producing 8 (rounded down).

The Date of month is self-explanatory. However, the Factor for month is an arbitrary number that you must memorize for each month, as shown below:

January:	1 (0 if leap year)	July:	0
February:	4 (3 if leap year)	August:	3
March:	4	September:	6
April:	0	October:	1
May:	2	November:	4
June:	5	December:	6

When you perform the division shown on the previous page, the day of the week is determined by the remainder, as shown below:

Remainder of 1 = Sunday
Remainder of 2 = Monday
Remainder of 3 = Tuesday
Remainder of 4 = Wednesday
Remainder of 5 = Thursday
Remainder of 6 = Friday
Remainder of 0 = Saturday

You may use these factors to determine the day of the week for any date, from 1900–1999. For any date within the years 2000–2099, you may apply the same information and formula, but you'll need to subtract 1 from the numerator of the formula.

You'll note above that the factors for January and February are diminished by 1 during leap years. During the twentieth century, the leap years were, or are, 1904, 1908, 1912, 1916, 1920, 1924, 1928, 1932, 1936, 1940, 1944, 1948, 1952, 1956, 1960, 1964, 1968, 1972, 1976, 1980, 1984, 1988, 1992, and 1996. The year 1900 was not a leap year. The year 2000 *is* a leap year, however, followed by leap years every four years throughout the twenty-first century.

Let's take an example—November 22, 1963. Apply the formula and you get $(63 + \frac{63}{4} + 22 + 4)/7$. The second number, $\frac{63}{4}$, is rounded down to 15, so the numerator totals 104. Now divide 7 into 104, and obtain 14 Remainder 6. All that matters is the remainder, so (according to the equivalencies above) November 22, 1963 fell on a Friday.

See if you can determine each day of the week in under 30 seconds:

A. May 14, 1952	E. Apr. 27, 1918	I. Mar. 9, 1986
B. Aug. 3, 1920	F. July 20, 1939	J. Nov. 30, 1967
C. Dec. 8, 1943	G. Feb. 11, 1977	K. June 12, 1900
D. Jan. 1, 1996	H. Oct. 18, 1909	L. Sept. 29, 1954

(See solutions on page 224)

Trick 14: Rapidly Check Multiplication and Division

Strategy: Let's finish week one by learning how to check multiplication and division with a technique called "casting out nines." When a calculation has been performed correctly, this trick will indicate as such. When the wrong answer has been obtained, this method will *probably*, but not definitely, uncover the error. To check multiplication, the basic idea is to first obtain "digit sums" for the multiplicand and multiplier. For example, the digit sum of 25 is 7 (2 + 5). **Only a one-digit digit sum can be used.** Therefore, whenever a digit sum exceeds nine, simply perform another digit-sum calculation. For example, the digit sum of 683 is 17 (6 + 8 + 3). However, we must then compute the digit sum of 17, which is 8 (1 + 7). The second step is to multiply the digit sums of the multiplicand and multiplier together, to obtain a third digit sum. If this third digit sum matches the digit sum of the computed answer, chances are that the answer is correct. If they do not match, then the answer is definitely incorrect. To save even more time, disregard totals of "9" along the way. The following examples will clarify this fascinating technique.

Elementary Example #1

31 × 11 = 341

```
   31 (digit sum = 4)
  ×11 (digit sum = 2)
 ____ (multiply: 4 × 2 = 8)
  341 (digit sum = 8)
```

Because the third and fourth digit sums are identical, the answer is probably correct.

Elementary Example #2

82 × 69 = 5,737

```
    83 (digit sum = 11; digit sum of 11 = 2)
   ×69 (digit sum = 15; digit sum of 15 = 6)
 _____ (multiply: 2 × 6 = 12; digit sum of 12 = 3)
 5,737 (digit sum = 22; digit sum of 22 = 4)
```

Because the third and fourth digit sums (3 and 4) do not agree, the answer is definitely incorrect. (Note: To save time, the 9 of the multiplier could have been disregarded, to obtain the same digit sum of 6.)

Brain Builder #1

836 × 794 = 663,784

```
     836 (digit sum = 17; digit sum of 17 = 8)
    ×794 (digit sum = 20; digit sum of 20 = 2)
 _______ (multiply: 8 × 2 = 16; digit sum of 16 = 7)
 663,784 (digit sum = 34; digit sum of 34 = 7)
```

Because the third and fourth digit sums are identical, the answer is probably correct.

Number-Power Note: This trick has been illustrated for multiplication only. However, because division is the inverse of multiplication, to apply the technique, simply convert the division into a multiplication problem. For example, to test 884 ÷ 26 = 34, look at it as 34 × 26 = 884, and go from there. This trick will also work with decimal points and zeroes—just disregard them. (They have been omitted here to simplify the explanation.) Trick 42 uses the same method, with a slight variation, to check addition and subtraction.

Remember, it is possible for the digit sums to agree but for the answer to be wrong. For example, in Elementary Example #1, an incorrect answer of 431 would produce the same digit sum, 8.

Elementary Exercises

Check the calculations below for accuracy, using the "casting out nines" method. For each one, indicate "probably correct" or "definitely incorrect." Reread the Number-Power Note to see how to check division problems for accuracy.

1. $53 \times 27 = 1,441$
2. $77 \times 22 = 1,694$
3. $96 \times 18 = 1,728$
4. $45 \times 600 = 277,000$
5. $14 \times 62 = 858$
6. $88 \times 33 = 2,904$
7. $71 \times 49 = 3,459$
8. $65 \times 24 = 1,560$
9. $38 \times 92 = 3,496$
10. $42 \times 56 = 2,352$
11. $5,846 \div 79 = 84$
12. $1,360 \div 85 = 16$
13. $3,149 \div 47 = 77$
14. $5,684 \div 98 = 58$
15. $2,349 \div 29 = 71$
16. $988 \div 13 = 76$

Brain Builders

1. $364 \times 826 = 300{,}664$
2. $555 \times 444 = 247{,}420$
3. $797 \times 51 = 40{,}647$
4. $286 \times 972 = 277{,}992$
5. $319 \times 634 = 202{,}276$
6. $740 \times 561 = 414{,}140$
7. $493 \times 168 = 82{,}824$
8. $666 \times 425 = 283{,}050$
9. $2,691{,}837 \div 857 = 3{,}141$
10. $992{,}070 \div 365 = 2{,}618$
11. $877{,}982 \div 217 = 4{,}046$
12. $689{,}976 \div 777 = 888$

(See solutions on page 203)

Week 1 Quick Quiz

Let's see how many tricks from week one you can remember and apply by taking this brief test. There's no time limit, but try to work through these items as rapidly as possible. Before you begin, glance at the computations and try to identify the trick that you could use. When you flip ahead to the solutions, you will see which trick was intended.

Elementary Examples

1. $45 \times 4 =$
2. $44 \div 5 =$
3. $11 \times 36 =$
4. $1{,}800 \div 60 =$
5. $900 \times 60 =$
6. $72 \times 25 =$
7. $99 \times 65 =$
8. $65^2 =$
9. $31 \times 29 =$
10. $76 \div 4 =$
11. $52 \times 5 =$
12. $93 \times 101 =$
13. $0.3 \times 700 =$
14. $640 \div 1.6 =$
15. $700 \div 25 =$
16. Using the "casting out nines" method, indicate whether this calculation is probably correct or definitely incorrect:
    $14 \times 87 = 1{,}318$

Brain Builders

1. $500 \times 4.7 =$
2. $350 \times 3.5 =$
3. $110 \times 5.8 =$
4. $580 \div 40 =$
5. $710 \div 500 =$
6. $211 \div 2.5 =$
7. $52 \times 40 =$
8. $3.9 \times 41 =$
9. $50 \times 9.4 =$
10. $7.5 \times 75 =$
11. $2.5 \times 36 =$
12. $1{,}010 \times 0.37 =$
13. $9.9 \times 0.85 =$
14. Using the "casting out nines" method, indicate whether this calculation is probably correct or definitely incorrect:
    $364{,}231 \div 853 = 427$

(See solutions on page 221)

Week 2 Multiplication and Division II

Trick 15: Rapidly Multiply by 125 (or 0.125, 1.25, 12.5, 1,250, etc.)

Strategy: We begin Week 2 with more excellent reciprocal techniques. To multiply a number by 125, **divide the number by 8,** and affix or insert any necessary zeroes or decimal point. The examples below illustrate how terrific this trick really is.

Elementary Example #1

8 × 125

Step 1. Divide: 8 ÷ 8 = 1 (intermediary quotient).

Step 2. Apply T of R: A quick estimate puts the answer at about 1,000.

Step 3. Affix three zeroes to the intermediary quotient, producing the answer 1,000.

Thought Process Summary

$$\begin{array}{r} 8 \\ \times 125 \\ \hline \end{array} \quad \rightarrow \quad 8 \div 8 = 1 \quad \rightarrow \quad 1{,}000$$

Elementary Example #2

32 × 125

Step 1. Divide: 32 ÷ 8 = 4 (intermediary quotient).

Step 2. Apply T of R: A quick estimate puts the answer between 3,000 and 5,000.

Step 3. Affix three zeroes to the intermediary quotient, producing the answer 4,000.

Thought Process Summary

32 × 125 → 32 ÷ 8 = 4 → 4,000

Brain Builder #1

120 × 1.25

Step 1. Disregard the zero and decimal point and think, "12 × 125."
Step 2. Divide: 12 ÷ 8 = 1.5 (intermediary quotient).
Step 3. Apply T of R: A quick estimate puts the answer in the mid-100s.
Step 4. Eliminate the decimal point from, and affix one zero to, the intermediary quotient, producing the answer 150.

Thought Process Summary

120 × 1.25 → 12 × 125 → 12 ÷ 8 = 1.5 → 150

Brain Builder #2

7.2 × 12.5

Step 1. Disregard the decimal points, converting the problem to 72 × 125.
Step 2. Divide: 72 ÷ 8 = 9 (intermediary quotient).
Step 3. Apply T of R: A quick estimate puts the answer at just under 100.
Step 4. Affix one zero to the intermediary quotient, producing the answer 90.

Thought Process Summary

7.2 × 12.5 → 72 × 125 → 72 ÷ 8 = 9 → 90

Number-Power Note: Remember that quick estimates are best determined by rounding off the numbers involved. For example, in Elementary Example #2 (32 × 125), the multiplicand could be rounded down to 30, and to offset the rounding down, the multiplier could be rounded *up* to 130, producing an estimate of 3,900.

Elementary Exercises

These exercises are simple—when you know the trick!

1. $16 \times 125 =$
2. $40 \times 125 =$
3. $88 \times 125 =$
4. $56 \times 125 =$
5. $125 \times 24 =$
6. $125 \times 64 =$
7. $125 \times 96 =$
8. $125 \times 32 =$
9. $80 \times 125 =$
10. $160 \times 125 =$
11. $72 \times 125 =$
12. $48 \times 125 =$
13. $125 \times 120 =$
14. $125 \times 104 =$
15. $125 \times 8 =$
16. $125 \times 240 =$

Brain Builders

1. $20 \times 125 =$
2. $60 \times 125 =$
3. $36 \times 125 =$
4. $28 \times 125 =$
5. $1.25 \times 64 =$
6. $12.5 \times 3.2 =$
7. $1{,}250 \times 0.88 =$
8. $125 \times 0.96 =$
9. $560 \times 1.25 =$
10. $4.8 \times 125 =$

(See solutions on page 203)

Trick 16: Rapidly Divide by 125 (or 0.125, 1.25, 12.5, 1,250, etc.)

Strategy: You've probably already figured out this trick's shortcut. To divide a number by 125, **multiply the number by 8**, and affix or insert any necessary zeroes or decimal point. As you'll see in the following examples, dividing by 125 is even easier than multiplying by 125.

Elementary Example #1

300 ÷ 125

Step 1. Disregard the zeroes and think, "3 ÷ 125."

Step 2. Multiply: $3 \times 8 = 24$ (intermediary product).

Step 3. Apply T of R: A quick estimate puts the answer between 2 and 3.

Step 4. Insert a decimal point within the intermediary product, producing the answer 2.4.

Thought Process Summary

300 ÷ 125 → 3 ÷ 125 → 3 × 8 = 24 → 2.4

Elementary Example #2

111 ÷ 125

Step 1. Multiply: 111 × 8 = 888 (intermediary product).

Step 2. Apply T of R: A quick estimate puts the answer at just under 1.

Step 3. Affix a decimal point to the left of the intermediary product, producing the answer 0.888 (or just .888).

Thought Process Summary

111 ÷ 125 → 111 × 8 = 888 → 0.888

Brain Builder #1

70 ÷ 12.5

Step 1. Disregard the zero and decimal point and think, "7 ÷ 125."

Step 2. Multiply: 7 × 8 = 56 (intermediary product).

Step 3. Apply T of R: A quick estimate puts the answer between 5 and 6.

Step 4. Insert a decimal point within the intermediary product, producing the answer 5.6.

Thought Process Summary

70 ÷ 12.5 → 7 ÷ 125 → 7 × 8 = 56 → 5.6

Brain Builder #2

45 ÷ 1.25

Step 1. Disregard the decimal point and think, "45 ÷ 125."

Step 2. Multiply: 45 × 8 = 360 (intermediary product).

Step 3. Apply T of R: A quick estimate puts the answer between 30 and 40.

Step 4. Insert a decimal point within the intermediary product, producing the answer 36.

Thought Process Summary

$45 \div 1.25 \rightarrow 45 \div 125 \rightarrow 45 \times 8 = 360 \rightarrow 36$

Number-Power Note: When obtaining a quick estimate, it isn't necessary to come really close to the true answer. The only purpose of the estimate is to direct you to the proper placement of the decimal point or to the correct number of zeroes to affix. For example, in Brain Builder #2, as long as you know that the answer is going to consist of two digits, you know that the intermediary product of 360 must be reduced to 36.

Elementary Exercises

As you work through these exercises, you'll see that multiplying by 8 is a whole lot easier than dividing by 125!

1. 200 ÷ 125 =
2. 6,000 ÷ 125 =
3. 1, 100 ÷ 125 =
4. 800 ÷ 125 =
5. 4,000 ÷ 125 =
6. 9,000 ÷ 125 =
7. 10,000 ÷ 125 =
8. 700 ÷ 125 =
9. 3,000 ÷ 125 =
10. 5,000 ÷ 125 =
11. 1,200 ÷ 125 =
12. 1,110 ÷ 125 =
13. 600 ÷ 125 =
14. 1,500 ÷ 125 =
15. 250 ÷ 125 =
16. 2,000 ÷ 125 =

Brain Builders

1. 40 ÷ 1.25 =
2. 90 ÷ 12.5 =
3. 750 ÷ 125 =
4. 11.1 ÷ 12.5 =
5. 2 ÷ 1.25 =
6. 60 ÷ 125 =
7. 350 ÷ 125 =
8. 7 ÷ 0.125 =
9. 50 ÷ 12.5 =
10. 2.5 ÷ 1.25 =

(See solutions on page 204)

Trick 17: Rapidly Multiply by 9 (or 0.9, 90, 900, etc.)

Strategy: We begin Day 9 appropriately with a trick involving the number 9. To multiply a number by 9, first **multiply the number by 10**. From that product, **subtract the number itself** to obtain the answer. Be sure to disregard any zeroes or decimals initially, and to reaffix or reinsert them, if necessary, to complete the calculation. Let's take a look at a few examples.

Elementary Example #1

14 × 9

Step 1. Multiply: 14 × 10 = 140.

Step 2. Subtract: 140 − 14 = 126 (the answer).

Thought Process Summary

$$\begin{array}{r} 14 \\ \times 9 \\ \hline \end{array} \rightarrow \begin{array}{r} 14 \\ \times 10 \\ \hline 140 \end{array} \rightarrow \begin{array}{r} 140 \\ -14 \\ \hline 126 \end{array}$$

Elementary Example #2

26 × 9

Step 1. Multiply: 26 × 10 = 260.

Step 2. Subtract: 260 − 26 = 234 (the answer).

Thought Process Summary

$$\begin{array}{r} 26 \\ \times 9 \\ \hline \end{array} \rightarrow \begin{array}{r} 26 \\ \times 10 \\ \hline 260 \end{array} \rightarrow \begin{array}{r} 260 \\ -26 \\ \hline 234 \end{array}$$

Brain Builder #1

450 × 0.9

Step 1. Disregard the zero and decimal point and think, "45 × 9."

Step 2. Multiply: 45 × 10 = 450.

Step 3. Subtract: 450 − 45 = 405 (intermediary difference).

Step 4. Apply T of R: Because 0.9 is just under 1, the answer must be just under 450. The intermediary difference of 405 is therefore the answer.

Thought Process Summary

$$\begin{array}{r} 450 \\ \times 0.9 \\ \hline \end{array} \rightarrow \begin{array}{r} 45 \\ \times 9 \\ \hline \end{array} \rightarrow \begin{array}{r} 45 \\ \times 10 \\ \hline 450 \end{array} \rightarrow \begin{array}{r} 450 \\ -45 \\ \hline 405 \end{array}$$

Brain Builder #2

7.5 × 900

Step 1. Disregard the decimal point and zeroes and think, "75 × 9."

Step 2. Multiply: 75 × 10 = 750.

Step 3. Subtract: 750 − 75 = 675 (intermediary difference).

Step 4. Apply T of R: A quick estimate puts the answer somewhere between 6,000 and 7,000.

Step 5. Affix one zero to the intermediary difference, producing the answer 6,750.

Thought Process Summary

$$\begin{array}{r} 7.5 \\ \times 900 \\ \hline \end{array} \rightarrow \begin{array}{r} 75 \\ \times 9 \\ \hline \end{array} \rightarrow \begin{array}{r} 75 \\ \times 10 \\ \hline 750 \end{array} \rightarrow \begin{array}{r} 750 \\ -75 \\ \hline 675 \end{array} \rightarrow 6{,}750$$

Number-Power Note: This trick is more difficult to apply than most of the others, because you must perform mental subtraction—a feat that is far more difficult than mental addition. However, with practice, and with the completion of the rapid subtraction section of this book, you will become adept at multiplying by 9.

Elementary Exercises

You'll need to apply an extra dose of concentration when doing these exercises.

1. $13 \times 9 =$
2. $24 \times 9 =$
3. $35 \times 9 =$
4. $17 \times 9 =$
5. $9 \times 12 =$
6. $9 \times 25 =$
7. $9 \times 55 =$
8. $9 \times 15 =$
9. $28 \times 9 =$
10. $67 \times 9 =$
11. $34 \times 9 =$
12. $16 \times 9 =$
13. $9 \times 56 =$
14. $9 \times 18 =$
15. $9 \times 19 =$
16. $9 \times 23 =$

Brain Builders

1. $7.8 \times 9 =$
2. $89 \times 90 =$
3. $270 \times 0.9 =$
4. $0.29 \times 900 =$
5. $90 \times 3.8 =$
6. $9 \times 95 =$
7. $900 \times 0.47 =$
8. $0.09 \times 690 =$
9. $360 \times 90 =$
10. $0.58 \times 900 =$

(See solutions on page 204)

Mathematical Curiosity #3

$142{,}857 \times 1 = 142{,}857$
$142{,}857 \times 2 = 285{,}714$ (same digits, rearranged)
$142{,}857 \times 3 = 428{,}571$ (rearranged yet another way)
$142{,}857 \times 4 = 571{,}428$ (rearranged yet another way)
$142{,}857 \times 5 = 714{,}285$ (rearranged yet another way)
$142{,}857 \times 6 = 857{,}142$ (rearranged yet another way)
$142{,}857 \times 7 = 999{,}999$ (here's where the pattern ends)

(Also note: $142 + 857 = 999$, and $14 + 28 + 57 = 99$)

Trick 18: Rapidly Multiply by 12 (or 0.12, 1.2, 120, etc.)

Strategy: This trick is especially useful when dealing with dozens. To rapidly multiply a number by 12, first **multiply the number by 10**. To that product, **add twice the number** to obtain the answer. Be sure to disregard any zeroes or decimals initially, and to reaffix or reinsert them, if necessary, to complete the calculation. The following exercises illustrate a variety of applications.

Elementary Example #1

25 × 12

Step 1. Multiply: 25 × 10 = 250.

Step 2. Multiply: 25 × 2 = 50.

Step 3. Add: 250 + 50 = 300 (the answer).

Thought Process Summary

$$\begin{array}{r} 25 \\ \times 12 \\ \hline \end{array} \rightarrow \begin{array}{r} 25 \\ \times 10 \\ \hline 250 \end{array} \rightarrow \begin{array}{r} 25 \\ \times 2 \\ \hline 50 \end{array} \rightarrow \begin{array}{r} 250 \\ +50 \\ \hline 300 \end{array}$$

Elementary Example #2

33 × 12

Step 1. Multiply: 33 × 10 = 330.

Step 2. Multiply: 33 × 2 = 66.

Step 3. Add: 330 + 66 = 396 (the answer).

Thought Process Summary

$$\begin{array}{r} 33 \\ \times 12 \\ \hline \end{array} \rightarrow \begin{array}{r} 33 \\ \times 10 \\ \hline 330 \end{array} \rightarrow \begin{array}{r} 33 \\ \times 2 \\ \hline 66 \end{array} \rightarrow \begin{array}{r} 330 \\ +66 \\ \hline 396 \end{array}$$

Brain Builder #1

650 × 0.12

Step 1. Disregard the zero and decimal point and think, "65 × 12."

Step 2. Multiply: 65 × 10 = 650.

Step 3. Multiply: 65 × 2 = 130.

Step 4. Add: 650 + 130 = 780 (intermediary sum).

Step 5. Apply T of R: 650 × 0.1 would equal 65. Therefore, 650 × 0.12 should equal slightly more than 65.

Step 6. Insert a decimal point within the intermediary sum, producing the answer 78.

Thought Process Summary

$$\begin{array}{r}650\\ \times 0.12\\ \hline\end{array} \rightarrow \begin{array}{r}65\\ \times 12\\ \hline\end{array} \rightarrow \begin{array}{r}65\\ \times 10\\ \hline 650\end{array} \rightarrow \begin{array}{r}65\\ \times 2\\ \hline 130\end{array} \rightarrow \begin{array}{r}650\\ +130\\ \hline 780\end{array} \rightarrow 78$$

Brain Builder #2

1.9 × 1.2

Step 1. Disregard the decimal points and think, "19 × 12."

Step 2. Multiply: 19 × 10 = 190.

Step 3. Multiply: 19 × 2 = 38.

Step 4. Add: 190 + 38 = 228 (intermediary sum).

Step 5. Apply T of R: 2 × 1.2 would equal 2.4. Therefore, 1.9 × 1.2 should equal slightly less than 2.4.

Step 6. Insert a decimal point within the intermediary sum, producing the answer 2.28.

Thought Process Summary

$$\begin{array}{r}1.9\\ \times 1.2\\ \hline\end{array} \rightarrow \begin{array}{r}19\\ \times 12\\ \hline\end{array} \rightarrow \begin{array}{r}19\\ \times 10\\ \hline 190\end{array} \rightarrow \begin{array}{r}19\\ \times 2\\ \hline 38\end{array} \rightarrow \begin{array}{r}190\\ +38\\ \hline 228\end{array} \rightarrow 2.28$$

Number-Power Note: This trick requires a bit more concentration than most, because you must independently perform two multiplications and then add the products. With practice, however, you will be able to add it to your repertoire.

Elementary Exercises

Pretend that your brain is the memory button on a calculator when working these exercises.

1. 45 × 12 =
2. 18 × 12 =
3. 16 × 12 =
4. 75 × 12 =
5. 12 × 22 =
6. 12 × 14 =
7. 12 × 15 =
8. 12 × 55 =
9. 32 × 12 =
10. 21 × 12 =
11. 17 × 12 =
12. 35 × 12 =
13. 12 × 24 =
14. 12 × 85 =
15. 12 × 31 =
16. 12 × 23 =

Brain Builders

1. 340 × 1.2 =
2. 1.7 × 120 =
3. 5.5 × 12 =
4. 230 × 12 =
5. 120 × 0.95 =
6. 0.12 × 1,500 =
7. 1.2 × 3.2 =
8. 12 × 190 =
9. 220 × 0.12 =
10. 0.85 × 120 =

(See solutions on page 204)

Trick 19: Rapidly Multiply by 15 (or 0.15, 1.5, 150, etc.)

Strategy: This trick is especially useful in computing a 15-percent restaurant tip. To multiply a number by 15, first **multiply the number by 10**. To that product, **add half of the product** to obtain the answer. Be sure to disregard any zeroes or decimal points initially, and to reaffix or reinsert them, if necessary, to complete the calculation. Let's check out some examples of this practical trick.

Elementary Example #1

12 × 15

Step 1. Multiply: 12 × 10 = 120.

Step 2. Halve: 120 ÷ 2 = 60.

Step 3. Add: 120 + 60 = 180 (the answer). (*Note:* Why not use Trick 18 to check your answer?)

Thought Process Summary

$$\begin{array}{r} 12 \\ \times 15 \\ \hline \end{array} \rightarrow \begin{array}{r} 12 \\ \times 10 \\ \hline 120 \end{array} \rightarrow 120 \div 2 = 60 \rightarrow \begin{array}{r} 120 \\ +60 \\ \hline 180 \end{array}$$

Elementary Example #2

66 × 15

Step 1. Multiply: 66 × 10 = 660.

Step 2. Halve: 660 ÷ 2 = 330.

Step 3. Add: 660 + 330 = 990 (the answer).

Thought Process Summary

$$\begin{array}{r} 66 \\ \times 15 \\ \hline \\ \end{array} \rightarrow \begin{array}{r} 66 \\ \times 10 \\ \hline 660 \end{array} \rightarrow 660 \div 2 = 330 \rightarrow \begin{array}{r} 660 \\ +330 \\ \hline 990 \end{array}$$

Brain Builder #1

$28 × 15%

Step 1. Multiply: $28 × 10% = $2.80.

Step 2. Halve: $2.80 ÷ 2 = $1.40.

Step 3. Add: $2.80 + $1.40 = $4.20 (the answer).

Thought Process Summary

$$\begin{array}{r} \$28 \\ \times 15\% \\ \hline \\ \end{array} \rightarrow \begin{array}{r} \$28 \\ \times 10\% \\ \hline \$2.80 \end{array} \rightarrow \$2.80 \div 2 = \$1.40 \rightarrow \begin{array}{r} \$2.80 \\ +1.40 \\ \hline \$4.20 \end{array}$$

Brain Builder #2

0.32 × 150

Step 1. Disregard the decimal point and zero and think, "32 × 15."

Step 2. Multiply: 32 × 10 = 320.

Step 3. Halve: 320 ÷ 2 = 160.

Step 4. Add: 320 + 160 = 480 (intermediary sum).

Step 5. Apply T of R: 0.32 is approximately $\frac{1}{3}$. Therefore, 0.32 × 150 is around $\frac{1}{3}$ of 150, or about 50.

Step 6. Insert a decimal point within the intermediary sum, producing the answer 48.0 (or just 48).

Thought Process Summary

$$\begin{array}{r} 0.32 \\ \times 150 \\ \hline \\ \end{array} \rightarrow \begin{array}{r} 32 \\ \times 15 \\ \hline \\ \end{array} \rightarrow \begin{array}{r} 32 \\ \times 10 \\ \hline 320 \end{array} \rightarrow 320 \div 2 = 160 \rightarrow \begin{array}{r} 320 \\ +160 \\ \hline 480 \end{array} \rightarrow 48$$

Number-Power Note: Instead of multiplying by 10, halving, and then adding the two amounts, you could halve, add the two amounts, and *then* multiply by 10. For example, in Elementary Example #2, you could start with 66, add 33 (producing 99), and *then* multiply the 99 by 10 to produce the answer 990. Choose whichever variation is more to your liking.

Elementary Exercises

Don't let the dollar and percentage signs bother you as you work through these exercises.

1. $8 × 15% =
2. $16 × 15% =
3. 44 × 15 =
4. 26 × 15 =
5. 15% × $14 =
6. 15% × $38 =
7. 15 × 34 =
8. 15 × 42 =
9. $18 × 15% =
10. $22 × 15% =
11. 62 × 15 =
12. 54 × 15 =
13. $24 × 15% =
14. $58 × 15% =
15. 15 × 46 =
16. 15 × 52 =

Brain Builders

1. 88 × 15 =
2. 72 × 15 =
3. $5.60 × 15% =
4. $4.80 × 15% =
5. 15 × 360 =
6. 1.5 × 160 =
7. $6.40 × 15% =
8. $68 × 15% =
9. 180 × 1.5 =
10. 25 × 15 =

(See solutions on page 205)

Number Potpourri #2

Do you know the names of the numbers, as used in the United States, starting with one thousand, and increasing one thousandfold each time? Answer: Thousand, million, billion, trillion, quadrillion, quintillion, sextillion, septillion, octillion, nonillion, decillion, and so on.

Trick 20: Rapidly Multiply Two Numbers with a Special Relationship—the Tens Digits Are the Same and the Ones Digits Add Up to Ten

Strategy: This trick is a variation on Trick 7, which you might want to review before reading further. An example of a calculation that fits the above description is 23 × 27. Note that the tens digits are both 2, and that the ones digits add up to 10 (3 + 7). Using 23 × 27 as an example, **multiply the tens digit** (2) **by the next whole number** (3), to obtain a product of 6. Then, **multiply the ones digits together** (3 × 7) to obtain a product of 21. Finally, combine the two products to obtain the answer 621. Let's go over some examples of this most unusual trick.

Elementary Example #1

16 × 14

Step 1. Multiply: 1 × 2 = 2.

Step 2. Multiply: 6 × 4 = 24.

Step 3. Combine: 224 (the answer). (Why not use Trick 13 to check this answer?)

Thought Process Summary

$$\begin{array}{r} 16 \\ \times 14 \\ \hline \end{array} \rightarrow \begin{array}{r} 1 \\ \times 2 \\ \hline 2 \end{array} \rightarrow \begin{array}{r} 6 \\ \times 4 \\ \hline 24 \end{array} \rightarrow 224$$

Elementary Example #2

71 × 79

Step 1. Multiply: 7 × 8 = 56.

Step 2. Multiply: 1 × 9 = 09 (Note the zero affixed to the one-digit product).

Step 3. Combine: 5,609 (the answer).

Thought Process Summary

$$\begin{array}{r} 71 \\ \times 79 \\ \hline \end{array} \rightarrow \begin{array}{r} 7 \\ \times 8 \\ \hline 56 \end{array} \rightarrow \begin{array}{r} 1 \\ \times 9 \\ \hline 09 \end{array} \rightarrow 5{,}609$$

Brain Builder #1

6.2 × 6.8

Step 1. Disregard the decimal points and think, "62 × 68."

Step 2. Multiply: 6 × 7 = 42.

Step 3. Multiply: 2 × 8 = 16.

Step 4. Combine: 4,216 (intermediary product).

Step 5. Apply T of R: Obtain a quick estimate by rounding (for example, 6 × 7 = 42).

Step 6. Insert a decimal point within the intermediary product, producing the answer 42.16.

Thought Process Summary

$$\begin{array}{r}6.2\\ \times 6.8\\ \hline\end{array} \rightarrow \begin{array}{r}62\\ \times 68\\ \hline\end{array} \rightarrow \begin{array}{r}6\\ \times 7\\ \hline 42\end{array} \rightarrow \begin{array}{r}2\\ \times 8\\ \hline 16\end{array} \rightarrow 4,216 \rightarrow 42.16$$

Brain Builder #2

470 × 0.43

Step 1. Disregard the zero and decimal point, and think, "47 × 43."

Step 2. Multiply: 4 × 5 = 20.

Step 3. Multiply: 7 × 3 = 21.

Step 4. Combine: 2,021 (intermediary product).

Step 5. Apply T of R: A quick estimate puts the answer at about 200.

Step 6. Insert a decimal point within the intermediary product, producing the answer 202.1.

Thought Process Summary

$$\begin{array}{r}470\\ \times 0.43\\ \hline\end{array} \rightarrow \begin{array}{r}47\\ \times 43\\ \hline\end{array} \rightarrow \begin{array}{r}4\\ \times 5\\ \hline 20\end{array} \rightarrow \begin{array}{r}7\\ \times 3\\ \hline 21\end{array} \rightarrow 2,021 \rightarrow 202.1$$

Number-Power Note: There is also a technique to multiply two numbers whose *tens* digits add to 10, and whose *ones* digits are the same. However, the technique is cumbersome and its use is not as easily identified as that in Trick 20, so it will not be included.

Elementary Exercises

It's hard to believe that you can perform these seemingly unrelated exercises with the same trick.

1. 18 × 12 =
2. 33 × 37 =
3. 84 × 86 =
4. 29 × 21 =
5. 58 × 52 =
6. 97 × 93 =
7. 24 × 26 =
8. 81 × 89 =
9. 56 × 54 =
10. 77 × 73 =
11. 39 × 31 =
12. 64 × 66 =
13. 71 × 79 =
14. 99 × 91 =
15. 42 × 48 =
16. 88 × 82 =

Brain Builders

1. 440 × 4.6 =
2. 190 × 11 =
3. 7.4 × 7.6 =
4. 5.9 × 51 =
5. 2.8 × 220 =
6. 0.98 × 920 =
7. 360 × 3.4 =
8. 13 × 170 =
9. 0.67 × 63 =
10. 0.94 × 0.96 =

(See solutions on page 205)

Trick 21: Rapidly Multiply by 1.5, 2.5, 3.5, etc.

Strategy: This trick assumes that it is easier to work with whole numbers than with halves. To rapidly multiply by 1.5, 2.5, and so forth, **double the 1.5, 2.5, or other multiplier,** and **halve the other number,** to (hopefully) convert the computation into one of whole numbers only. In fact, this technique will work when multiplying by *any* number ending in 5. Here are some good examples using this clever trick.

Elementary Example #1

3.5 × 12

Step 1. Double: $3.5 \times 2 = 7$.

Step 2. Halve: $12 \div 2 = 6$.

Step 3. Multiply: $7 \times 6 = 42$ (the answer).

Thought Process Summary

$$\begin{array}{r} 3.5 \\ \times 12 \\ \hline \end{array} \rightarrow \begin{array}{r} 3.5 \\ \times 2 \\ \hline 7 \end{array} \rightarrow 12 \div 2 = 6 \rightarrow \begin{array}{r} 7 \\ \times 6 \\ \hline 42 \end{array}$$

Elementary Example #2

4.5 × 16

Step 1. Double: $4.5 \times 2 = 9$.

Step 2. Halve: $16 \div 2 = 8$.

Step 3. Multiply: $9 \times 8 = 72$ (the answer).

Thought Process Summary

$$\begin{array}{r}4.5\\ \times 16\\ \hline\end{array} \rightarrow \begin{array}{r}4.5\\ \times 2\\ \hline 9\end{array} \rightarrow 16 \div 2 = 8 \rightarrow \begin{array}{r}9\\ \times 8\\ \hline 72\end{array}$$

Brain Builder #1

85 × 22

Step 1. Double: 85 × 2 = 170.

Step 2. Halve: 22 ÷ 2 = 11.

Step 3. Multiply: 170 × 11 = 1,870 (the answer). (Did you remember to use Trick 8—the "11 trick"—for Step 3?)

Thought Process Summary

$$\begin{array}{r}85\\ \times 22\\ \hline\end{array} \rightarrow \begin{array}{r}85\\ \times 2\\ \hline 170\end{array} \rightarrow 22 \div 2 = 11 \rightarrow \begin{array}{r}170\\ \times 11\\ \hline 1,870\end{array}$$

Brain Builder #2

7.5 × 320

Step 1. Disregard the decimal point and zero, and think, "75 × 32."

Step 2. Double: 75 × 2 = 150.

Step 3. Halve: 32 ÷ 2 = 16.

Step 4. Multiply: 150 × 16 = 2,400 (intermediary product). (Did you remember to use Trick 19—the "15 trick"—for Step 4?)

Step 5. Apply T of R: A quick estimate puts the answer somewhere in the 2,000s. The intermediary product of 2,400 is also the answer.

Thought Process Summary

$$\begin{array}{r}7.5\\ \times 320\\ \hline\end{array} \rightarrow \begin{array}{r}75\\ \times 32\\ \hline\end{array} \rightarrow \begin{array}{r}75\\ \times 2\\ \hline 150\end{array} \rightarrow 32 \div 2 = 16 \rightarrow \begin{array}{r}150\\ \times 16\\ \hline 2,400\end{array}$$

Number-Power Note: Try this trick the next time you are in a restaurant and must calculate a 15-percent tip. Also, in Brain Builder #2, you could apply the same technique to Step 4 by converting 150 × 16 into 300 × 8 (which is an easier way to arrive at 2,400).

Elementary Exercises

Just think "Double and Halve" when working through these exercises.

1. 14 × 1.5 =
2. 22 × 7.5 =
3. 126 × 5.5 =
4. 8 × 9.5 =
5. 6.5 × 12 =
6. 8.5 × 6 =
7. 2.5 × 18 =
8. 3.5 × 24 =
9. 26 × 7.5 =
10. 48 × 1.5 =
11. 46 × 5.5 =
12. 8 × 6.5 =
13. 4.5 × 18 =
14. 3.5 × 28 =
15. 9.5 × 6 =
16. 2.5 × 34 =

Brain Builders

1. 600 × 0.85 =
2. 24 × 35 =
3. 1.44 × 55 =
4. 80 × 6.5 =
5. 150 × 4.8 =
6. 950 × 0.8 =
7. 75 × 7.2 =
8. 2,500 × 0.22 =
9. 160 × 4.5 =
10. 80 × 8.5 =

(See solutions on page 205)

Trick 22: Rapidly Divide by 1.5, 2.5, 3.5, etc.

Strategy: As you may have guessed, this trick is very similar to Trick 21. To divide by 1.5, 2.5, or the like, **double both the divisor and the dividend** to (probably) convert the computation into one of whole numbers only. In fact, this technique will work when dividing by *any* number ending in 5. You'll realize the simplicity of this trick after you've gone over the following examples.

Elementary Example #1

28 ÷ 3.5

Step 1. Double: 28 × 2 = 56.

Step 2. Double: 3.5 × 2 = 7.

Step 3. Divide: 56 ÷ 7 = 8 (the answer).

Thought Process Summary

$$28 \div 3.5 \rightarrow \begin{array}{r} 28 \\ \times 2 \\ \hline 56 \end{array} \rightarrow \begin{array}{r} 3.5 \\ \times 2 \\ \hline 7 \end{array} \rightarrow 56 \div 7 = 8$$

Elementary Example #2

26 ÷ 6.5

Step 1. Double: 26 × 2 = 52.

Step 2. Double: 6.5 × 2 = 13.

Step 3. Divide: 52 ÷ 13 = 4 (the answer).

Thought Process Summary

$$26 \div 6.5 \rightarrow \begin{array}{r} 26 \\ \times 2 \\ \hline 52 \end{array} \rightarrow \begin{array}{r} 6.5 \\ \times 2 \\ \hline 13 \end{array} \rightarrow 52 \div 13 = 4$$

Brain Builder #1

225 ÷ 45

Step 1. Double: 225 × 2 = 450.

Step 2. Double: 45 × 2 = 90.

Step 3. Divide: 450 ÷ 90 = 5 (the answer).

Thought Process Summary

$$225 \div 45 \rightarrow \begin{array}{r} 225 \\ \times 2 \\ \hline 450 \end{array} \rightarrow \begin{array}{r} 45 \\ \times 2 \\ \hline 90 \end{array} \rightarrow 450 \div 90 = 5$$

Brain Builder #2

315 ÷ 10.5

Step 1. Double: 315 × 2 = 630.

Step 2. Double: 10.5 × 2 = 21.

Step 3. Divide: 630 ÷ 21 = 30 (the answer).

Thought Process Summary

$$315 \div 10.5 \rightarrow \begin{array}{r}315\\ \times 2\\ \hline 630\end{array} \rightarrow \begin{array}{r}10.5\\ \times 2\\ \hline 21\end{array} \rightarrow 630 \div 21 = 30$$

Number-Power Note: Try Trick 22 for the following practical application: You have driven 375 miles since your last tank fill-up. How many miles per gallon did you realize, assuming that you just filled up the tank with 15 gallons?

Elementary Exercises

Just think, "Double, double, and divide" when working through these exercises.

1. $33 \div 5.5 =$
2. $34 \div 8.5 =$
3. $22.5 \div 2.5 =$
4. $52 \div 6.5 =$
5. $37.5 \div 7.5 =$
6. $24.5 \div 3.5 =$
7. $27 \div 4.5 =$
8. $38 \div 9.5 =$
9. $12 \div 1.5 =$
10. $39 \div 6.5 =$
11. $14 \div 3.5 =$
12. $170 \div 8.5 =$
13. $225 \div 7.5 =$
14. $44 \div 5.5 =$
15. $28.5 \div 9.5 =$
16. $21 \div 1.5 =$

Brain Builders

1. $49 \div 3.5 =$
2. $175 \div 2.5 =$
3. $1{,}950 \div 65 =$
4. $1.05 \div 0.15 =$
5. $4{,}500 \div 750 =$
6. $475 \div 95 =$
7. $385 \div 55 =$
8. $180 \div 4.5 =$
9. $255 \div 8.5 =$
10. $24.5 \div 3.5 =$

(See solutions on page 206)

Trick 23: Rapidly Square Any Two-Digit Number Beginning in 5

Strategy: Day 12 covers the two remaining squaring techniques. To square any two-digit number beginning in 5, first **add 25 to the ones digit**. To that sum, **affix the square of the ones digit** (for 1, 2, and 3 squared, write "01," "04," and "09," respectively). You now have your answer. The number that must be added to the ones digit (25) is easy to remember because $5^2 = 25$. Although a calculation such as 530×5.3 is technically not a square, it *can* be solved using this technique. Here are some examples for you to peruse.

Elementary Example #1

51^2

Step 1. Add: $25 + 1 = 26$.
Step 2. Square: $1^2 = 01$.
Step 3. Combine: 2,601 (the answer).

Thought Process Summary

$$\begin{array}{r} 51 \\ \times 51 \\ \hline \end{array} \rightarrow \begin{array}{r} 25 \\ +1 \\ \hline 26 \end{array} \rightarrow \begin{array}{r} 1 \\ \times 1 \\ \hline 01 \end{array} \rightarrow 2{,}601$$

Elementary Example #2

58^2

Step 1. Add: $25 + 8 = 33$.
Step 2. Square: $8^2 = 64$.
Step 3. Combine: 3,364 (the answer).

Thought Process Summary

$$\begin{array}{r} 58 \\ \times 58 \\ \hline \end{array} \rightarrow \begin{array}{r} 25 \\ +8 \\ \hline 33 \end{array} \rightarrow \begin{array}{r} 8 \\ \times 8 \\ \hline 64 \end{array} \rightarrow 3{,}364$$

Brain Builder #1

5.4^2

Step 1. Disregard the decimal point and think, "54^2."

Step 2. Add: $25 + 4 = 29$.

Step 3. Square: $4^2 = 16$.

Step 4. Combine: 2,916 (intermediary product).

Step 5. Apply T of R: A quick estimate puts the answer near 30.

Step 6. Insert a decimal point within the intermediary product, producing the answer 29.16.

Thought Process Summary

$$\begin{array}{r} 5.4 \\ \times 5.4 \\ \hline \end{array} \rightarrow \begin{array}{r} 54 \\ \times 54 \\ \hline \end{array} \rightarrow \begin{array}{r} 25 \\ +4 \\ \hline 29 \end{array} \rightarrow \begin{array}{r} 4 \\ \times 4 \\ \hline 16 \end{array} \rightarrow 2{,}916 \rightarrow 29.16$$

Brain Builder #2

530×5.3

Step 1. Disregard the zero and decimal point and think, "53^2."

Step 2. Add: $25 + 3 = 28$.

Step 3. Square: $3^2 = 09$.

Step 4. Combine: 2,809 (intermediary product).

Step 5. Apply T of R: A quick estimate puts the answer in the 2,000's. Therefore, the intermediary product of 2,809 is the answer.

Thought Process Summary

$$\begin{array}{r} 530 \\ \times 5.3 \\ \hline \end{array} \rightarrow \begin{array}{r} 53 \\ \times 53 \\ \hline \end{array} \rightarrow \begin{array}{r} 25 \\ +3 \\ \hline 28 \end{array} \rightarrow \begin{array}{r} 3 \\ \times 3 \\ \hline 09 \end{array} \rightarrow 2{,}809$$

Number-Power Note: When you square a number that ends in zeroes, for every zero disregarded upon starting the calculation, remember to affix *two* zeroes upon completing the problem. For example, $560 \times 560 = 313,600$.

Elementary Exercises

Master yet another trick with these exercises:

1. $52^2 =$
2. $56 \times 56 =$
3. $59 \times 59 =$
4. $53^2 =$
5. $57 \times 57 =$
6. $55 \times 55 =$
7. $54^2 =$
8. $58 \times 58 =$
9. $51 \times 51 =$
10. $59^2 =$
11. $56^2 =$
12. $52 \times 52 =$
13. $53 \times 53 =$
14. $57^2 =$
15. $55^2 =$
16. $51^2 =$

Brain Builders

1. $5.7^2 =$
2. $510 \times 0.51 =$
3. $55 \times 5.5 =$
4. $0.52 \times 520 =$
5. $590 \times 0.59 =$
6. $56 \times 5.6 =$
7. $5.8 \times 5.8 =$
8. $53 \times 530 =$
9. $540^2 =$
10. $5.9 \times 59 =$

(See solutions on page 206)

Number Potpourri #3

Think fast! Add the following numbers out loud, in order, and as quickly as you can:

$$1,000 + 40 + 1,000 + 30 + 1,000 + 20 + 1,000 + 10 =$$

(See solution on page 225)

Trick 24: Rapidly Square Any Two-Digit Number Ending in 1

Strategy: This trick assumes that it is easier to square a number ending in zero than to square a number ending in one. It is also important to know that the difference between the squares of two consecutive numbers is equal to their sum. For example, $31^2 - 30^2 = 30 + 31$, and $71^2 - 70^2 = 70 + 71$. Using the same mathematical logic, to obtain 31^2, simply add $30^2 + 30 + 31$, which equals 961. Although a calculation such as 810×8.1 is technically not a square, it, too, can be solved using this trick. The following examples illustrate how easily this trick can be applied.

Elementary Example #1

21^2

Step 1. Square: $20^2 = 400$.

Step 2. Add: $20 + 21 = 41$.

Step 3. Add: $400 + 41 = 441$ (the answer).

Thought Process Summary

$$\begin{array}{r} 21 \\ \times 21 \\ \hline \end{array} \rightarrow \begin{array}{r} 20 \\ \times 20 \\ \hline 400 \end{array} \rightarrow \begin{array}{r} 20 \\ +21 \\ \hline 41 \end{array} \rightarrow \begin{array}{r} 400 \\ +41 \\ \hline 441 \end{array}$$

Elementary Example #2

51^2

Step 1. Square: $50^2 = 2{,}500$.

Step 2. Add: $50 + 51 = 101$.

Step 3. Add: $2{,}500 + 101 = 2{,}601$ (the answer). (Why not use Trick 23 to check this answer?)

Thought Process Summary

$$\begin{array}{r} 51 \\ \times 51 \\ \hline \end{array} \rightarrow \begin{array}{r} 50 \\ \times 50 \\ \hline 2{,}500 \end{array} \rightarrow \begin{array}{r} 50 \\ +51 \\ \hline 101 \end{array} \rightarrow \begin{array}{r} 2{,}500 \\ +101 \\ \hline 2{,}601 \end{array}$$

Brain Builder #1

810 × 8.1

Step 1. Disregard the zero and decimal point and think, "81^2."
Step 2. Square: $80^2 = 6{,}400$.
Step 3. Add: $80 + 81 = 161$.
Step 4. Add: $6{,}400 + 161 = 6{,}561$ (intermediary product).
Step 5. Apply T of R: A quick estimate puts the answer in the 6,000s. Therefore, the intermediary product of 6,561 is the answer.

Thought Process Summary

$$\begin{array}{r} 810 \\ \times 8.1 \\ \hline \end{array} \rightarrow \begin{array}{r} 81 \\ \times 81 \\ \hline \end{array} \rightarrow \begin{array}{r} 80 \\ \times 80 \\ \hline 6{,}400 \end{array} \rightarrow \begin{array}{r} 80 \\ +81 \\ \hline 161 \end{array} \rightarrow \begin{array}{r} 6{,}400 \\ +161 \\ \hline 6{,}561 \end{array}$$

Brain Builder #2

410^2

Step 1. Disregard the zero and think, "41^2."
Step 2. Square: $40^2 = 1{,}600$.
Step 3. Add: $40 + 41 = 81$.
Step 4. Add: $1{,}600 + 81 = 1{,}681$ (intermediary product).
Step 5. Apply T of R: Two zeroes must be affixed to the intermediary product (because one was initially disregarded), producing the answer 168,100.

Thought Process Summary

$$\begin{array}{r} 410 \\ \times 410 \\ \hline \end{array} \rightarrow \begin{array}{r} 41 \\ \times 41 \\ \hline \end{array} \rightarrow \begin{array}{r} 40 \\ \times 40 \\ \hline 1{,}600 \end{array} \rightarrow \begin{array}{r} 40 \\ +41 \\ \hline 81 \end{array} \rightarrow \begin{array}{r} 1{,}600 \\ +81 \\ \hline 1{,}681 \end{array} \rightarrow 168{,}100$$

Number-Power Note: This trick would also work for the squaring of numbers ending in 9, although the two numbers would have to be subtracted, not added (for example, $29^2 = 30^2 - 30 - 29 = 841$). However, performing the subtraction mentally is, for the most part, very cumbersome.

Elementary Exercises

These exercises will help you reinforce this extremely useful trick.

1. $31 \times 31 =$
2. $61 \times 61 =$
3. $91 \times 91 =$
4. $11 \times 11 =$
5. $71 \times 71 =$
6. $41 \times 41 =$
7. $81 \times 81 =$
8. $51 \times 51 =$
9. $21^2 =$
10. $31^2 =$
11. $61^2 =$
12. $71^2 =$
13. $41^2 =$
14. $91^2 =$
15. $51^2 =$
16. $81^2 =$

Brain Builders

1. $210 \times 2.1 =$
2. $5.1^2 =$
3. $7.1 \times 71 =$
4. $0.91 \times 910 =$
5. $310^2 =$
6. $61 \times 6.1 =$
7. $4.1 \times 410 =$
8. $81 \times 0.81 =$
9. $0.51 \times 510 =$
10. $21 \times 210 =$

(See solutions on page 206)

Trick 25: Rapidly Multiply Two-Digit Numbers Without Showing Work

Strategy: This trick never fails to impress others when pulled off. To multiply two 2-digit numbers without showing work, first **multiply the ones digits together**, then "**cross-multiply**," and finally **multiply the tens digits together**. Make sure to carry whenever a product exceeds 9. As you will see in the examples below, you must work your way from right to left to perform this trick.

Elementary Example #1

12 × 23

Step 1. Multiply the ones digits: $2 \times 3 = 6$ (ones-digit answer).

Step 2. Cross-multiply and add: $(1 \times 3) + (2 \times 2) = 7$ (tens-digit answer).

Step 3. Multiply the tens digits: $1 \times 2 = 2$ (hundreds-digit answer).

Step 4. Combine: 276 (the answer).

Thought Process Summary

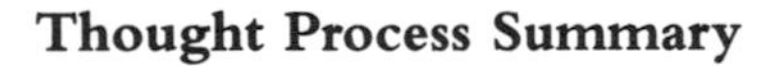

$$\begin{array}{r} 12 \\ \times 23 \\ \hline \end{array} \rightarrow \begin{array}{r} 12 \\ \times 23 \\ \hline 6 \end{array} \rightarrow \begin{array}{r} 12 \\ \times 23 \\ \hline 76 \end{array} \rightarrow \begin{array}{r} 12 \\ \times 23 \\ \hline 276 \end{array}$$

Elementary Example #2

31 × 24

Step 1. Multiply the ones digits: $1 \times 4 = 4$ (ones-digit answer).

Step 2. Cross-multiply and add: $(3 \times 4) + (1 \times 2) = 14$ (use the 4 as the tens-digit answer, and carry the 1).

Step 3. Multiply the tens digits, and carry the 1: $(3 \times 2) + 1 = 7$ (hundreds-digit answer).

Step 4. Combine: 744 (the answer).

Thought Process Summary

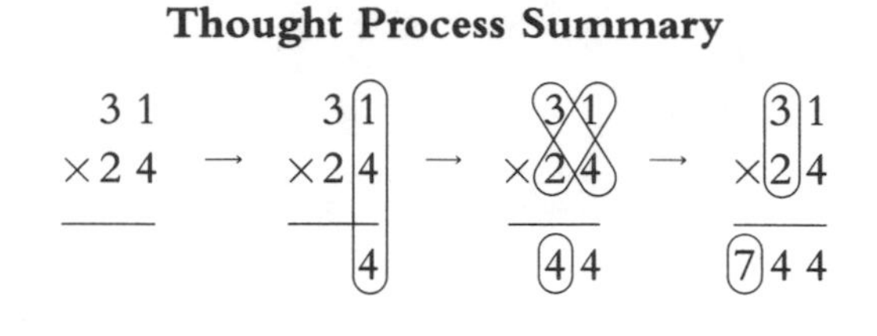

Brain Builder #1

76 × 54

Step 1. Multiply the ones digits: $6 \times 4 = 24$ (use the 4 as the ones digit answer, and carry the 2).

Step 2. Cross-multiply, add, and carry the 2: $(6 \times 5) + (7 \times 4) + 2 = 60$ (use the 0 as the tens-digit answer, and carry the 6).

Step 3. Multiply the tens digits, and carry the 6: $(7 \times 5) + 6 = 41$ (the thousands- and hundreds-digit answers).

Step 4. Combine: 4,104 (the answer).

Thought Process Summary

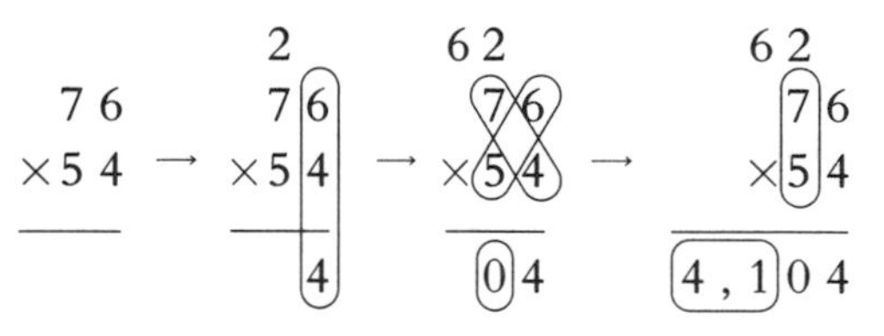

Brain Builder #2

67 × 98

Step 1. Multiply the ones digits: $7 \times 8 = 56$ (use the 6 as the ones-digit answer, and carry the 5).

Step 2. Cross-multiply, add, and carry the 5: $(7 \times 9) + (6 \times 8) + 5 = 116$ (use the 6 as the tens-digit answer, and carry the 11).

Step 3. Multiply the tens digits, and carry the 11: $(6 \times 9) + 11 = 65$ (the thousands- and hundreds-digit answers).

Step 4. Combine: 6,566 (the answer).

Thought Process Summary

$$\begin{array}{r} 67 \\ \times 98 \\ \hline \end{array} \rightarrow \begin{array}{r} 5 \\ 67 \\ \times 98 \\ \hline 6 \end{array} \rightarrow \begin{array}{r} 115 \\ 67 \\ \times 98 \\ \hline 66 \end{array} \rightarrow \begin{array}{r} 115 \\ 67 \\ \times 98 \\ \hline 6,566 \end{array}$$

Number-Power Note: This trick will also work on numbers with zeroes and decimal points. However, they have been omitted here to simplify the explanation. Also, as you are cross-multiplying, it is generally easier to begin with the larger product and then add the smaller product (see Trick 39 for a more complete description). Parlor Trick #3 on page 86 is an advanced variation of this trick.

Elementary Exercises

Amaze your friends and relatives by just writing down the answers to these exercises!

1. $\begin{array}{r} 13 \\ \times 21 \\ \hline \end{array}$

2. $\begin{array}{r} 22 \\ \times 14 \\ \hline \end{array}$

3. $\begin{array}{r} 32 \\ \times 15 \\ \hline \end{array}$

4. $\begin{array}{r} 42 \\ \times 42 \\ \hline \end{array}$

5. $\begin{array}{r} 51 \\ \times 33 \\ \hline \end{array}$

6. $\begin{array}{r} 25 \\ \times 13 \\ \hline \end{array}$

7. $\begin{array}{r} 44 \\ \times 11 \\ \hline \end{array}$

8. $\begin{array}{r} 53 \\ \times 34 \\ \hline \end{array}$

9. $\begin{array}{r} 14 \\ \times 52 \\ \hline \end{array}$

10. $\begin{array}{r} 43 \\ \times 45 \\ \hline \end{array}$

11. $\begin{array}{r} 35 \\ \times 41 \\ \hline \end{array}$

12. $\begin{array}{r} 23 \\ \times 55 \\ \hline \end{array}$

13. $\begin{array}{r} 54 \\ \times 15 \\ \hline \end{array}$

14. $\begin{array}{r} 21 \\ \times 51 \\ \hline \end{array}$

15. $\begin{array}{r} 34 \\ \times 34 \\ \hline \end{array}$

16. $\begin{array}{r} 12 \\ \times 45 \\ \hline \end{array}$

Brain Builders

1. $\begin{array}{r} 14 \\ \times 36 \\ \hline \end{array}$

2. $\begin{array}{r} 66 \\ \times 44 \\ \hline \end{array}$

3. $\begin{array}{r} 76 \\ \times 54 \\ \hline \end{array}$

4. $\begin{array}{r} 92 \\ \times 87 \\ \hline \end{array}$

5. $\begin{array}{r} 88 \\ \times 63 \\ \hline \end{array}$

6. $\begin{array}{r} 58 \\ \times 85 \\ \hline \end{array}$

7. $\begin{array}{r} 72 \\ \times 49 \\ \hline \end{array}$

8. $\begin{array}{r} 27 \\ \times 51 \\ \hline \end{array}$

9. $\begin{array}{r} 99 \\ \times 32 \\ \hline \end{array}$

10. $\begin{array}{r} 38 \\ \times 81 \\ \hline \end{array}$

(See solutions on page 207)

Parlor Trick #3: Multiply Two 3-Digit Numbers Without Showing Work

Strategy: This trick is performed the same way as Trick 25 (multiplying two-digit numbers without showing work), but carries the technique one step further. First you multiply the ones digits together, then cross-multiply three times, and finally multiply the hundreds digits together. Make sure to carry whenever a product exceeds 9. (This parlor trick requires a lot of concentration and practice, but never fails to amaze. If you really want a challenge, see if you can figure out how to multiply two 4-digit numbers without showing work!)

Demonstration Problem A

321 × 213

Thought Process Summary

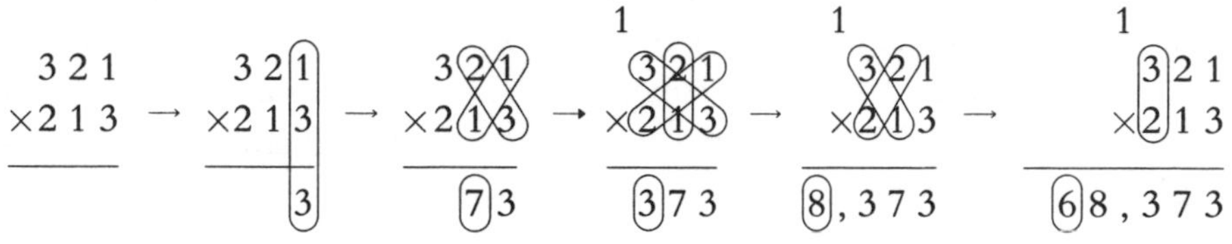

321 × 213 → 3 → 73 → 373 (carry 1) → 8,373 (carry 1) → 68,373

Mult.	Cross-mult.	Cross-mult.	Cross-mult.	Mult.
1 × 3	2 × 3	3 × 3	3 × 1	3 × 2
(3)	1 × 1	2 × 1	2 × 2	(6)
	Sum = 7	1 × 2	+1 carried	
		Sum = 13	Sum = 8	

Demonstration Problem B

263 × 417

Thought Process Summary

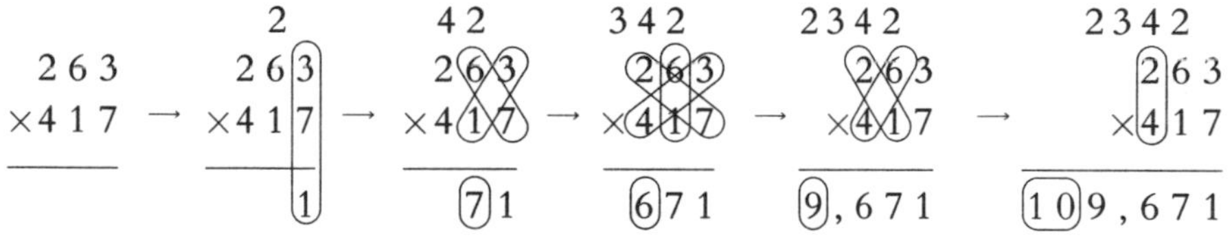

263 × 417 → 1 (carry 2) → 71 (carry 4 2) → 671 (carry 3 4 2) → 9,671 (carry 2 3 4 2) → 109,671

Mult.	Cross-mult.	Cross-mult.	Cross-mult.	Mult.
3 × 7	6 × 7	2 × 7	2 × 1	2 × 4
(21)	3 × 1	6 × 1	6 × 4	+2 carried
	+2 carried	3 × 4	+3 carried	Sum = 10
	Sum = 47	+4 carried	Sum = 29	
		Sum = 36		

Amaze your friends and yourself with these:

A	B	C	D	E	F
425 ×132	204 ×315	153 ×240	442 ×351	524 ×933	151 ×434

(See solutions on page 224)

Trick 26: Rapidly Multiply Two Numbers Whose Difference Is 4

Strategy: Here is another trick that tests your ability to square numbers. To multiply two numbers whose difference is 4, first **square the number that is exactly in the middle of the two**. Then **subtract 4 from that product** to obtain the answer. It's easy to remember the number to subtract (4) because it's the same as the difference between the two numbers to multiply. Continue reading for some demonstrations of this interesting trick.

Elementary Example #1

8 × 12

Step 1. Square the number in between: $10^2 = 100$.

Step 2. Subtract 4: $100 - 4 = 96$ (the answer).

Thought Process Summary

$$\begin{array}{r} 8 \\ \times 12 \\ \hline \end{array} \rightarrow \begin{array}{r} 10 \\ \times 10 \\ \hline 100 \end{array} \rightarrow \begin{array}{r} 100 \\ -4 \\ \hline 96 \end{array}$$

Elementary Example #2

13 × 9

Step 1. Square the number in between: $11^2 = 121$.

Step 2. Subtract 4: $121 - 4 = 117$ (the answer).

Thought Process Summary

$$\begin{array}{r} 13 \\ \times 9 \\ \hline \end{array} \rightarrow \begin{array}{r} 11 \\ \times 11 \\ \hline 121 \end{array} \rightarrow \begin{array}{r} 121 \\ -4 \\ \hline 117 \end{array}$$

Brain Builder #1

1.8 × 2.2

Step 1. Disregard the decimal points and think, "18 × 22."

Step 2. Square the number in between: $20^2 = 400$.

Step 3. Subtract 4: 400 − 4 = 396 (intermediary difference).

Step 4. Apply T of R: A quick estimate puts the answer near 4.

Step 5. Insert a decimal point within the intermediary product, producing the answer 3.96.

Thought Process Summary

1.8		18		20		400		
×2.2	→	×22	→	×20	→	−4	→	3.96
				400		396		

Brain Builder #2

1.7 × 130

Step 1. Disregard the decimal point and zero and think, "17 × 13."

Step 2. Square the number in between: $15^2 = 225$.

Step 3. Subtract 4: 225 − 4 = 221 (intermediary difference).

Step 4. Apply T of R: A quick estimate puts the answer in the 200s. The intermediary product of 221 is therefore the answer. (Note: Why not use Trick 20—"multiply two numbers with a special relationship"—to check this answer?)

Thought Process Summary

1.7		17		15		225
×130	→	×13	→	×15	→	−4
				225		221

Number-Power Note: This trick is a variation on Trick 13 (Rapidly Multiply Two Numbers Whose Difference is 2). Other tricks of this sort do exist. For example, if the difference between the numbers to multiply is 10, square the number in the middle and subtract 25. Or if the difference is 20, square the number in the middle and subtract 100. The two applications chosen for inclusion in this book are probably the easiest to remember, as well as the most useful.

Elementary Exercises

Some of these exercises can also be completed using previously learned tricks.

1. $11 \times 15 =$
2. $13 \times 17 =$
3. $33 \times 29 =$
4. $12 \times 8 =$
5. $9 \times 13 =$
6. $16 \times 12 =$
7. $28 \times 32 =$
8. $47 \times 43 =$
9. $27 \times 23 =$
10. $23 \times 19 =$
11. $18 \times 14 =$
12. $43 \times 39 =$
13. $57 \times 53 =$
14. $102 \times 98 =$
15. $68 \times 72 =$
16. $49 \times 53 =$

Brain Builders

1. $9 \times 130 =$
2. $19 \times 15 =$
3. $17 \times 21 =$
4. $5.8 \times 6.2 =$
5. $630 \times 6.7 =$
6. $2.3 \times 19 =$
7. $0.88 \times 920 =$
8. $180 \times 1.4 =$
9. $7.3 \times 7.7 =$
10. $0.83 \times 790 =$

(See solutions on page 207)

Trick 27: Rapidly Multiply in Two Steps

Strategy: We begin Day 14 with a trick that's simple, but which few people ever think to use. When numbers seem just a bit too large to multiply comfortably, **divide one number into two smaller ones** to make the calculation more manageable. For example, multiplying 7 × 16 might be faster and easier when viewed as 7 × 8 × 2. It is assumed that the following examples are just beyond most people's reach.

Elementary Example #1

8 × 14

Step 1. Split 14 into two parts: 7 × 2.

Step 2. Restate problem: 8 × 7 × 2 = 56 × 2 = 112 (the answer).

Thought Process Summary

8 × 14 → 8 × 7 = 56 → 56 × 2 = 112

Elementary Example #2

9 × 18

Step 1. Split 18 into two parts: 9 × 2.

Step 2. Restate problem: 9 × 9 × 2 = 81 × 2 = 162 (the answer).

Thought Process Summary

9 × 18 → 9 × 9 = 81 → 81 × 2 = 162

Brain Builder #1

14 × 2.2

Step 1. Disregard the decimal point and think, "14 × 22."

Step 2. Split 22 into two parts: 11 × 2.

Step 3. Restate problem: 14 × 11 × 2 = 154 × 2 = 308 (intermediary product). (Note the use of the "11 trick" in step 3.)

Step 4. Apply T of R: A quick estimate puts the answer at about 30.

Step 5. Insert a decimal point within the intermediary product, producing the answer 30.8.

Thought Process Summary

$$\begin{array}{r} 14 \\ \times 2.2 \\ \hline \end{array} \rightarrow \begin{array}{r} 14 \\ \times 22 \\ \hline \end{array} \rightarrow \begin{array}{r} 14 \\ \times 11 \\ \hline 154 \end{array} \rightarrow \begin{array}{r} 154 \\ \times 2 \\ \hline 308 \end{array} \rightarrow 30.8$$

Brain Builder #2

16 × 320

Step 1. Disregard the zero and think, "16 × 32."

Step 2. Split 32 into two parts: 16 × 2.

Step 3. Restate the problem: 16 × 16 × 2 = 256 × 2 = 512.

Step 4. Apply T of R: The zero that was initially disregarded must be reaffixed, producing the answer 5,120.

Thought Process Summary

$$\begin{array}{r} 16 \\ \times 320 \\ \hline \end{array} \rightarrow \begin{array}{r} 16 \\ \times 32 \\ \hline \end{array} \rightarrow \begin{array}{r} 16 \\ \times 16 \\ \hline 256 \end{array} \rightarrow \begin{array}{r} 256 \\ \times 2 \\ \hline 512 \end{array} \rightarrow 5{,}120$$

Number-Power Note: There are actually many tricks that involve breaking a calculation into two or more parts. Trick 57, for example, will show you how to multiply 78 by 6, or 706 by 8. The possibilities for this technique are limited only by your imagination.

Elementary Exercises

When breaking these exercises down into smaller components, you'll often find that another trick will help complete the calculation.

1. $7 \times 14 =$
2. $9 \times 12 =$
3. $8 \times 16 =$
4. $6 \times 18 =$
5. $15 \times 22 =$
6. $5 \times 18 =$
7. $12 \times 24 =$
8. $13 \times 22 =$
9. $9 \times 16 =$
10. $9 \times 14 =$
11. $8 \times 18 =$
12. $17 \times 22 =$
13. $7 \times 18 =$
14. $6 \times 16 =$
15. $7 \times 16 =$
16. $5 \times 16 =$

Brain Builders

1. $6 \times 1.6 =$
2. $80 \times 1.8 =$
3. $1.7 \times 2.2 =$
4. $7 \times 160 =$
5. $90 \times 1.4 =$
6. $0.9 \times 16 =$
7. $60 \times 18 =$
8. $7 \times 1.8 =$
9. $130 \times 2.2 =$
10. $8 \times 160 =$

(See solutions on page 207)

Mathematical Curiosity #4

$$1^2 = 1$$
$$11^2 = 121$$
$$111^2 = 12321$$
$$1111^2 = 1234321$$
$$11111^2 = 123454321$$
$$111111^2 = 12345654321$$

etc.

Trick 28: Rapidly Multiply Two Numbers That Are Just Over 100

Strategy: Let's finish Week 2 with a trick that will work nicely for calculations such as (102×103) and (109×104). **The answer will always be a five-digit number beginning in 1. The next two digits will equal the sum of the**

ones digits, and the last two digits will equal their product. A sum or product of one digit, such as 6, is written as two digits, or 06 in this case. Let's look at some examples of this clever trick.

Elementary Example #1

102 × 103

Step 1. Begin the answer: 1.

Step 2. Add the ones digits: 2 + 3 = 05.

Step 3. Multiply the ones digits: 2 × 3 = 06.

Step 4. Combine, writing from left to right: 10,506 (the answer). (Note: The first three digits—(105)—could have been obtained by simply adding 3 to 102 or 2 to 103.)

Thought Process Summary

$$\begin{array}{r}102\\ \times 103\\ \hline\end{array} \rightarrow 1 \rightarrow \begin{array}{r}2\\ +3\\ \hline 05\end{array} \rightarrow \begin{array}{r}2\\ \times 3\\ \hline 06\end{array} \rightarrow 10{,}506$$

Elementary Example #2

109 × 104

Step 1. Begin the answer: 1.

Step 2. Add the ones digits: 9 + 4 = 13.

Step 3. Multiply the ones digits: 9 × 4 = 36.

Step 4. Combine, writing from left to right: 11,336 (the answer).

Thought Process Summary

$$\begin{array}{r}109\\ \times 104\\ \hline\end{array} \rightarrow 1 \rightarrow \begin{array}{r}9\\ +4\\ \hline 13\end{array} \rightarrow \begin{array}{r}9\\ \times 4\\ \hline 36\end{array} \rightarrow 11{,}336$$

This trick will also work for a calculation such as 107 × 112. In this case, however, it is the 7 and the 12 that must be added and multiplied to produce the answer, 11,984. (This trick would not work for a calculation such as 108 × 113, because the product of 8 and 13 exceeds 99.) Let's look at a couple more of these advanced applications.

Brain Builder #1

116 × 105

Step 1. Begin the answer: 1.

Step 2. Add the ones and tens digits: 16 + 5 = 21.

Step 3. Multiply the ones and tens digits: 16 × 5 = 80.

Step 4. Combine, writing from left to right: 12,180 (the answer).

Thought Process Summary

$$\begin{array}{r} 116 \\ \times 105 \\ \hline \end{array} \rightarrow 1 \rightarrow \begin{array}{r} 16 \\ +5 \\ \hline 21 \end{array} \rightarrow \begin{array}{r} 16 \\ \times 5 \\ \hline 80 \end{array} \rightarrow 12{,}180$$

Brain Builder #2

101 × 127

Step 1. Begin the answer: 1.

Step 2. Add the ones and tens digits: 1 + 27 = 28.

Step 3. Multiply the ones and tens digits: 1 × 27 = 27.

Step 4. Combine, writing from left to right: 12,827 (the answer).

Thought Process Summary

$$\begin{array}{r} 101 \\ \times 127 \\ \hline \end{array} \rightarrow 1 \rightarrow \begin{array}{r} 1 \\ +27 \\ \hline 28 \end{array} \rightarrow \begin{array}{r} 1 \\ \times 27 \\ \hline 27 \end{array} \rightarrow 12{,}827$$

Number-Power Note: Trick 28 will also work with decimals (such as 1.01 × 10.7); however, for purposes of simplification only whole numbers have been used in our illustrations.

There is also a trick for multiplying two numbers that are just *under* 100. However, because this trick is difficult to remember and apply, it has not been included in this book.

Elementary Exercises

When doing these exercises, remember to add, then multiply.

1. $101 \times 101 =$
2. $107 \times 105 =$
3. $103 \times 106 =$
4. $109 \times 109 =$
5. $104 \times 102 =$
6. $108 \times 107 =$
7. $101 \times 109 =$
8. $106 \times 106 =$
9. $105 \times 104 =$
10. $107 \times 101 =$
11. $108 \times 109 =$
12. $101 \times 105 =$
13. $103 \times 107 =$
14. $108 \times 106 =$
15. $109 \times 105 =$
16. $102 \times 108 =$

Brain Builders

1. $106 \times 112 =$
2. $115 \times 104 =$
3. $113 \times 107 =$
4. $109 \times 111 =$
5. $103 \times 125 =$
6. $133 \times 102 =$
7. $112 \times 108 =$
8. $105 \times 117 =$
9. $101 \times 167 =$
10. $122 \times 104 =$

(See solutions on page 208)

Week 2 Quick Quiz

Let's see how many tricks from Week 2 you can remember and apply by taking this brief test. There's no time limit, but try to work through these items as rapidly as possible. Before you begin, glance at the computations and try to identify the trick that you could use. When you flip ahead to the solutions, you will see which trick was intended.

Elementary Problems

1. $26 \times 9 =$
2. $67 \times 63 =$
3. $3.5 \times 18 =$
4. $18 \times 7 =$
5. $125 \times 56 =$
6. $52^2 =$
7. Compute without showing work:
   $$\begin{array}{r} 24 \\ \times 31 \\ \hline \end{array}$$
8. $12 \times 65 =$
9. $29 \times 21 =$
10. $44 \div 5.5 =$
11. $52 \times 48 =$
12. $31 \times 31 =$
13. $\$36 \times 15\% =$
14. $7,000 \div 125 =$
15. $58 \times 58 =$
16. $103 \times 108 =$

Brain Builders

1. $180 \times 1.2 =$
2. $300 \div 12.5 =$
3. $6.1 \times 610 =$
4. $7.8 \times 7.2 =$
5. $23 \times 1.9 =$
6. $1,800 \div 45 =$
7. $1.25 \times 640 =$
8. $118 \times 104 =$
9. $5.7 \times 57 =$
10. $64 \times 150 =$
11. $70 \times 1.6 =$
12. $360 \times 0.9 =$
13. $4.5 \times 160 =$
14. Compute without showing work:
    $$\begin{array}{r} 83 \\ \times 47 \\ \hline \end{array}$$

(See solutions on page 221)

Week 3 Addition and Subtraction

Trick 29: Rapidly Subtract by Adding

Strategy: We're temporarily going to leave multiplication and division to concentrate this week on addition and subtraction. Remember that addition is generally easier to perform than subtraction. Whenever possible, therefore, **you should solve a subtraction problem by adding.** For example, to subtract 27 from 50, you should think, "27 plus what will equal 50?" Furthermore, adding from left to right, or "all at once," will usually be faster than handling the ones digit and then the tens digit. Let's look at some examples.

Elementary Example #1

72 − 56

Step 1. Think, "56 plus what equals 72?"

Step 2. Then think, "56, 66, 72," keeping in mind that you have added 10, then 6, for an answer of 16.

(*Note:* In Step 2, the tens were added, and then the ones. It is preferable to do it that way, rather than the other way around.)

Elementary Example #2

62 − 35

Step 1. Think, "35 plus what equals 62?"

Step 2. Then think, "35, 45, 55, 62," keeping in mind that you have added 20, then 7, for an answer of 27.

Brain Builder #1

121 − 75

Step 1. Think, "75 plus what equals 121?"

Step 2. Then think, "75, 85, 95, 105, 115, 121," keeping in mind that you have added 40, then 6, for an answer of 46.

Brain Builder #2

137 − 85

Step 1. Think, "85 plus what equals 137?"

Step 2. Simply by inspection, and by adding left to right, you can determine that the answer is 52.

Number-Power Note: With practice, you'll be able to perform this type of computation in almost no time at all. For example, you will look at a problem such as Brain Builder #1, and just think, "46." At first, however, you may have to count on your fingers the number of tens added.

Elementary Exercises

As you work through these exercises, remember to subtract by adding.

1. 31 − 17
2. 45 − 28
3. 60 − 32
4. 93 − 68
5. 87 − 36
6. 54 − 28
7. 43 − 19
8. 78 − 25
9. 81 − 45
10. 36 − 17
11. 59 − 11
12. 90 − 53
13. 51 − 33
14. 74 − 37
15. 48 − 26
16. 63 − 18

Brain Builders

1. 107 − 82
2. 113 − 79
3. 120 − 77
4. 124 − 88
5. 105 − 47
6. 111 − 66
7. 136 − 77
8. 122 − 31
9. 119 − 44
10. 104 − 77

(See solutions on page 208)

Trick 30: Rapidly Subtract by Adding—A Variation

Strategy: Here is more evidence that, whenever possible, you should subtract by adding. When the minuend and subtrahend of a subtraction problem are on opposite sides of 100, 200, or the like, **determine the distance** (number) **each is from that multiple of 100**, and **add those two numbers together** to obtain the answer. The following examples will clarify this very useful technique.

Elementary Example #1

143 − 98

Step 1. Think, "143 is 43 above 100, and 98 is 2 below 100."

Step 2. Add: 43 + 2 = 45 (the answer).

Elementary Example #2

155 − 87

Step 1. Think, "155 is 55 above 100, and 87 is 13 below 100."

Step 2. Add: 55 + 13 = 68 (the answer).

Brain Builder #1

245 − 189

Step 1. Think, "145 is 45 above 200, and 189 is 11 below 200."

Step 2. Add: 45 + 11 = 56 (the answer).

Brain Builder #2

1,038 − 979

Step 1. Think, "1,038 is 38 above 1,000, and 979 is 21 below 1,000."

Step 2. Add: 38 + 21 = 59 (the answer).

Number-Power Note: This trick will also work when the numbers are on opposite sides of *different* multiples of 100. For example, Brain Builder#1 above would have only taken you a split second longer to compute if it had asked you to subtract 189 from 345 (just take 11 + 145).

Elementary Exercises

With practice, you'll be able to work exercises like these in almost no time!

1. $\begin{array}{r} 108 \\ -92 \\ \hline \end{array}$
2. $\begin{array}{r} 132 \\ -87 \\ \hline \end{array}$
3. $\begin{array}{r} 124 \\ -66 \\ \hline \end{array}$
4. $\begin{array}{r} 111 \\ -55 \\ \hline \end{array}$
5. $\begin{array}{r} 117 \\ -75 \\ \hline \end{array}$
6. $\begin{array}{r} 123 \\ -84 \\ \hline \end{array}$
7. $\begin{array}{r} 136 \\ -93 \\ \hline \end{array}$
8. $\begin{array}{r} 142 \\ -89 \\ \hline \end{array}$
9. $\begin{array}{r} 166 \\ -78 \\ \hline \end{array}$
10. $\begin{array}{r} 105 \\ -48 \\ \hline \end{array}$
11. $\begin{array}{r} 151 \\ -85 \\ \hline \end{array}$
12. $\begin{array}{r} 120 \\ -58 \\ \hline \end{array}$
13. $\begin{array}{r} 170 \\ -77 \\ \hline \end{array}$
14. $\begin{array}{r} 115 \\ -98 \\ \hline \end{array}$
15. $\begin{array}{r} 103 \\ -44 \\ \hline \end{array}$
16. $\begin{array}{r} 150 \\ -63 \\ \hline \end{array}$

Brain Builders

1. $\begin{array}{r} 302 \\ -275 \\ \hline \end{array}$
2. $\begin{array}{r} 431 \\ -375 \\ \hline \end{array}$
3. $\begin{array}{r} 213 \\ -187 \\ \hline \end{array}$
4. $\begin{array}{r} 630 \\ -577 \\ \hline \end{array}$
5. $\begin{array}{r} 528 \\ -495 \\ \hline \end{array}$
6. $\begin{array}{r} 806 \\ -758 \\ \hline \end{array}$
7. $\begin{array}{r} 740 \\ -663 \\ \hline \end{array}$
8. $\begin{array}{r} 1,007 \\ -971 \\ \hline \end{array}$
9. $\begin{array}{r} 929 \\ -880 \\ \hline \end{array}$
10. $\begin{array}{r} 322 \\ -256 \\ \hline \end{array}$

(See solutions on page 208)

Trick 31: Rapidly Subtract by Altering

Strategy: Congratulations! You have reached the midpoint in this number-power program. Keep up the good work. The theory behind today's trick is that multiples of 10 are easier to subtract than nonmultiples. In addition, the answer to a subtraction problem will not be altered if you **change both the minuend and subtrahend by the same amount**, and **in the same direction**. For example, 25 − 8 will produce the same answer as 27 − 10. You'll note that both the minuend and subtrahend were increased by 2, to produce a slightly easier computation. When the subtrahend (the number to subtract from the minuend) is just *under* a multiple of ten, for example 58, it is best to **add** to convert it to a multiple of 10, or 60 in this case. However, when the subtrahend is just *over* a multiple of 10, for example 31, it is best to **subtract** to convert it to a multiple of 10, or 30 in this case. As you'll see in a moment, this trick is not nearly as complicated as it sounds.

Elementary Example #1
86 − 59

Step 1. Add 1 to both numbers, and think, "87 − 60 = 27" (the answer).

Elementary Example #2
44 − 28

Step 1. Add 2 to both numbers, and think, "46 − 30 = 16" (the answer).

Brain Builder #1
61 − 32

Step 1. Deduct 2 from both numbers, and think, "59 − 30 = 29" (the answer).

Brain Builder #2

90 − 51

Step 1. Deduct 1 from both numbers, and think, "89 − 50 = 39" (the answer).

Brain Builder #3

630 − 485

Step 1. Add 15 to each number, and think, "645 − 500 = 145" (the answer).

Number-Power Note: When applying this trick, use your imagination. For example, how would you handle 2,200 − 775? That's right, add 225 to each number, to produce the easier problem 2,425 − 1,000, which equals 1,425.

Elementary Exercises

When working these exercises, concentrate on converting the subtrahend to a multiple of 10.

1. 64 − 39
2. 72 − 38
3. 55 − 18
4. 81 − 49
5. 47 − 29
6. 90 − 48
7. 73 − 38
8. 32 − 19
9. 66 − 39
10. 80 − 68
11. 61 − 28
12. 100 − 79
13. 96 − 58
14. 70 − 29
15. 76 − 48
16. 53 − 19

Brain Builders

1. 86 − 31
2. 60 − 22
3. 50 − 32
4. 91 − 62
5. 70 − 21
6. 101 − 52
7. 80 − 41
8. 240 − 189
9. 160 − 85
10. 420 − 388

(See solutions on page 209)

Mathematical Curiosity #5

$$12^2 = 144 \text{ while } 21^2 = 441$$
$$13^2 = 169 \text{ while } 31^2 = 961$$
$$112^2 = 12{,}544 \text{ while } 211^2 = 44{,}521$$

(For each pair, the number to be squared has been reversed, producing the reverse result.)

Trick 32: Rapidly Add by Altering

Strategy: The theory behind this trick is that multiples of 10 are easier to add than nonmultiples. It is very similar to Trick 31. When an addend is just under a multiple of ten, it is usually faster and easier to **round the number up, add,** and then **subtract the same number that was added when rounding**. For example, it would be faster to add 85 + 59 by thinking, "85 + 60 = 145," and then subtracting the 1 that you previously added to the 59, producing the answer 144. Let's take a look at a few examples.

Elementary Example #1
54 + 38

Step 1. Add 2 to the 38, and think, "54 + 40 = 94."

Step 2. Subtract 2 from the 94, producing the answer 92.

Elementary Example #2
88 + 55

Step 1. Add 2 to the 88, and think, "90 + 55 = 145."

Step 2. Subtract 2 from the 145, producing the answer 143.

Brain Builder #1
276 + 149

Step 1. Add 1 to the 149, and think, "276 + 150 = 426."

Step 2. Subtract 1 from the 426, producing the answer 425.

(*Note:* A variation would be to subtract 1 from the 276 as you are adding 1 to the 149, converting the problem to 275 + 150.)

Brain Builder #2

388 + 442

Step 1. Add 12 to the 388, and think, "400 + 442 = 842."

Step 2. Subtract 12 from the 842, producing the answer 830.

(*Note:* You could round the 388 to 390 (and then subtract 2), but adding 12 and rounding to 400 is probably more efficient.)

Number-Power Note: This trick would also work with an addend slightly *over* a multiple of ten (as in 85 + 62). However, in this case the calculation is easy enough without having to make alterations.

Elementary Exercises

Remember to add and then subtract when doing these exercises.

1. 49 + 25 =
2. 56 + 28 =
3. 54 + 68 =
4. 19 + 76 =
5. 89 + 45 =
6. 38 + 94 =
7. 33 + 79 =
8. 16 + 58 =
9. 98 + 65 =
10. 39 + 74 =
11. 36 + 69 =
12. 85 + 48 =
13. 29 + 96 =
14. 78 + 44 =
15. 63 + 59 =
16. 95 + 18 =

Brain Builders

1. 178 + 65 =
2. 49 + 127 =
3. 54 + 129 =
4. 45 + 189 =
5. 126 + 68 =
6. 59 + 114 =
7. 38 + 133 =
8. 188 + 75 =
9. 144 + 79 =
10. 48 + 86 =

(See solutions on page 209)

Trick 33: Rapidly Add by Grouping and Reordering

Strategy: We begin Day 17 with a trick that requires a combination of practice and imagination. When adding several one-digit numbers, you should **group the numbers in combinations of 10** whenever possible, **add slightly out of order** when appropriate, and **"see" two or three numbers as their sum**, as when you are performing rapid reading. The following examples will show you, step by step, how to apply this indispensable technique.

Elementary Example #1

3 + 8 + 7 + 1 + 1 + 4 + 5 + 6 + 2

Step 1. Look at the 3 and 8 and instantly "see" 11.

Step 2. "See" the 7, 1, and 1 as 9. 11 + 9 = 20.

Step 3. Next, note that the 4 and (skipping ahead) the 6 total 10. 20 + 10 = 30.

Step 4. The only numbers left are 5 and 2, which equal 7. 30 + 7 = 37 (the answer).

Thought Process Summary

Think, "11, 20, 30, 37."

Elementary Example #2

5 + 2 + 7 + 7 + 3 + 6 + 4 + 4

Step 1. Note that the first two numbers total 7, and that two more 7s follow. Therefore, think, "Three 7s equal 21."

Step 2. Next, note that the 3 and the 6 total 9, and that 21 + 9 = 30.

Step 3. Finally, instantly "see" the 4 and 4 as 8. 30 + 8 = 38 (the answer).

Thought Process Summary

Think, "7, 21, 30, 38."

Brain Builder #1

9 + 7 + 4 + 8 + 5 + 1 + 2

Step 1. Look at the 9 and the 7, and "see" 16.

Step 2. Add: 16 + 4 = 20.

Step 3. Note that the 8 and (skipping ahead) the 2 equal 10. 20 + 10 = 30. (Be careful not to skip *too* far ahead when adding.)

Step 4. Look at the 5 and 1 and "see" 6. 30 + 6 = 36 (the answer).

Thought Process Summary

Think, "16, 20, 30, 36."

Brain Builder #2

5 + 3 + 2 + 4 + 8 + 6 + 7

Step 1. Look at the 5, 3, and 2, and "see" 10.

Step 2. Note that the 4 and the 6 total 10. 10 + 10 = 20.

Step 3. "See" the 8 and the 7 as 15. 20 + 15 = 35 (the answer).

Thought Process Summary

Think, "10, 20, 35."

Number-Power Note: Obviously, there might be several different ways to approach the rapid addition of a group of numbers. The best way to master this technique is simply to practice as much as possible. If you are a golfer, watch how most people struggle just to add nine little numbers in their head. Then see if they can add the back-nine total to the front-nine total, without a side computation, to compute the 18-hole total.

Elementary Exercises

The objective here is not only to obtain the correct answer, but to do so as quickly as possible by applying the three rapid-addition concepts you just learned.

1.	5.	9.	13.
7	5	9	4
3	6	9	2
5	3	4	4
6	4	5	9
5	9	3	5
8	7	8	1
6	1	7	7
+4	+8	+5	+3

2.	6.	10.	14.
2	3	6	1
6	4	2	7
9	3	7	6
1	8	5	3
3	6	9	5
7	2	1	9
5	4	4	5
+4	+5	+4	+8

3.	7.	11.	15.
8	3	4	9
8	5	7	6
8	5	3	4
6	7	1	7
3	9	8	6
8	1	8	2
2	6	9	8
+7	+6	+5	+3

4.	8.	12.	16.
1	2	2	3
5	3	3	3
9	5	7	6
5	1	9	8
2	1	9	7
2	6	1	4
2	9	5	9
+7	+7	+6	+1

Brain Builders

1.	4.	7.	10.
8	5	9	8
8	8	3	5
5	2	7	6
3	7	4	4
7	5	2	9
1	1	6	3
9	9	1	7
+4	+6	+7	+7

2.	5.	8.
4	7	2
7	7	5
3	9	9
3	9	5
3	1	3
5	4	4
9	5	8
+2	+8	+1

3.	6.	9.
6	3	6
3	3	2
9	3	1
4	7	8
8	1	5
5	8	5
2	5	7
+2	+6	+3

(See solutions on page 209)

Number Potpourri #4

How many times more powerful is an 8.0 earthquake (on the Richter scale) than a 6.0? (see solution on page 225)

Trick 34: Rapidly Add Without Carrying

Strategy: When adding long columns of numbers, it is often cumbersome to have to "carry" from one column to the next. In addition, when long columns are involved, there is a high likelihood that an error will be made. Under this method, not only does one not have to carry, but also it will be easier to find an error if one has been made. Note in the examples below how each subtotal is moved one column to the left of the previous one.

Elementary Example #1

```
                           38
                           74
                           91
                           55
                           16
                           29
                           42
                          +83
                          ---
ones column total →        38
tens column total →       39
                          ---
           Answer →       428
```

Elementary Example #2

```
                           29
                           80
                           77
                           15
                           31
                           99
                           46
                          +67
                          ---
ones column total →        44
tens column total →       40
                          ---
           Answer →       444
```

Brain Builder #1

```
                           386
                           409
                            15
                           164
                           772
                           580
                            93
                          +929
                          ----
     ones column total →    38
     tens column total →   41
 hundreds column total →  2 9
                          ----
              Answer →   3, 348
```

Brain Builder#2

```
                             37
                            591
                            836
                          2,088
                            915
                          3,442
                          1,916
                        +   190
                        -------
     ones column total →     35
     tens column total →    38
 hundreds column total →   3 6
thousands column total →   6
                        -------
              Answer →   10,015
```

Number-Power Note: A variation on the above technique is to start with the left-most column, and work your way to the right. For example, in Brain Builder #1, under the "equals" symbol, you would write:

```
29
 41
  38
----
3,348
```

Elementary Exercises

Remember—carrying is *not* allowed when doing these exercises.

1.	5.	9.	13.
57	94	75	53
22	63	45	93
81	27	30	16
46	11	12	44
92	55	89	78
40	99	52	49
39	70	31	23
+74	+34	+58	+80

2.	6.	10.	14.
18	21	26	58
85	64	87	97
53	76	91	60
90	83	16	12
24	17	35	41
66	38	65	86
71	86	84	38
+47	+62	+23	+91

3.	7.	11.	15.
79	43	37	55
29	96	98	76
42	14	56	14
60	50	78	83
13	73	41	20
97	28	93	99
44	69	48	46
+33	+72	+95	+62

4.	8.	12.	16.
68	25	59	32
88	67	27	95
36	82	89	18
51	19	35	68
20	80	40	34
77	54	72	29
49	32	37	71
+15	+61	+66	+48

Brain Builders

1.
$$\begin{array}{r} 566 \\ 873 \\ 927 \\ 104 \\ 653 \\ 498 \\ 272 \\ +795 \\ \hline \end{array}$$

2.
$$\begin{array}{r} 416 \\ 123 \\ 945 \\ 808 \\ 391 \\ 728 \\ 576 \\ +252 \\ \hline \end{array}$$

3.
$$\begin{array}{r} 451 \\ 394 \\ 177 \\ 902 \\ 815 \\ 263 \\ 588 \\ +686 \\ \hline \end{array}$$

4.
$$\begin{array}{r} 833 \\ 244 \\ 197 \\ 713 \\ 406 \\ 924 \\ 392 \\ +597 \\ \hline \end{array}$$

5.
$$\begin{array}{r} 127 \\ 996 \\ 382 \\ 437 \\ 505 \\ 841 \\ 269 \\ +634 \\ \hline \end{array}$$

6.
$$\begin{array}{r} 559 \\ 284 \\ 339 \\ 414 \\ 987 \\ 165 \\ 789 \\ +837 \\ \hline \end{array}$$

7.
$$\begin{array}{r} 726 \\ 964 \\ 199 \\ 320 \\ 438 \\ 514 \\ 667 \\ +855 \\ \hline \end{array}$$

8.
$$\begin{array}{r} 174 \\ 717 \\ 966 \\ 324 \\ 409 \\ 753 \\ 858 \\ +681 \\ \hline \end{array}$$

9.
$$\begin{array}{r} 326 \\ 968 \\ 183 \\ 3,471 \\ 38 \\ 1,567 \\ 846 \\ +2,612 \\ \hline \end{array}$$

10.
$$\begin{array}{r} 132 \\ 2,548 \\ 456 \\ 1,035 \\ 261 \\ 91 \\ 4,878 \\ +762 \\ \hline \end{array}$$

(See solutions on page 210)

Trick 35: Rapidly Add a Column of Numbers—Variation 1

Strategy: The basic idea is to **add left to right**, either **number by number** (variation 1) or all tens digits and then all ones digits (variation 2, Trick 36). This method, which requires a bit more concentration than most, is best applied for numbers less than 100. In the following examples, the tens digit is added before the ones digit. In the Brain Builder examples, some additional shortcuts have been taken which can be applied with some practice.

Elementary Example #1

	Count quickly, as follows:
16	16
29	26, 36, 45
37	55, 65, 75, 82
34	92, 102, 112, 116
12	126, 128
+27	138, 148, 155 (the answer)

Elementary Example #2

	Count quickly, as follows:
34	34
26	44, 54, 60,
17	77 (It's faster to just add the entire number here.)
5	82
12	92, 94
+6	100 (the answer)

Brain Builder #1

	Count quickly, as follows:
73	73
14	83, 87
66	147, 153 (Jumping from 87 to 147 takes practice.)
38	183, 191
7	198
+25	218, 223 (the answer)

Brain Builder #2

	Count quickly, as follows:
11	11
75	81, 86 (Or, just look at 11 and 75 and think, "86.")
34	116, 120 (Jumping from 86 to 116 takes practice.)
68	188 (It's faster to just add the entire number here.)
49	228, 237
+26	257, 263 (the answer)

Number-Power Note: Being able to mentally add a column of numbers is one of the most useful number-power tools you can acquire. Choose either Variation 1 or Variation 2 (Trick 36), whichever is more to your liking.

Elementary Exercises

Remember to add left to right, number by number, when doing these exercises.

1.	3.	5.	7.
29	16	4	21
43	38	27	40
6	5	16	6
31	22	40	13
9	47	8	39
+17	+3	+13	+1

2.	4.	6.	8.
8	34	49	51
27	2	12	27
10	48	33	2
9	19	50	18
33	7	7	5
+42	+25	+25	+36

9. 14
37
9
23
18
+6

10. 40
4
28
17
11
+35

Brain Builders

1. 37
9
66
21
18
+54

2. 11
59
36
40
8
+23

3. 32
7
84
51
13
+49

4. 81
35
6
12
27
+44

5. 52
27
18
30
3
+46

6. 48
4
13
29
66
+31

7. 37
15
8
24
86
+41

8. 77
15
1
29
32
+50

9. 10
59
2
41
63
+14

10. 55
27
16
6
29
+5

(See solutions on page 210)

Trick 36: Rapidly Add a Column of Numbers—Variation 2

Strategy: This trick similar to Trick 35, except that **all the tens digits are added before the ones digits.** The same examples are presented here to emphasize the difference in execution. Note that it's faster to add *up* the ones column (rather than down) after the tens column has been completed. In effect, you are making a giant "U."

Elementary Example #1

Count quickly, as follows:

16	10		155 (the answer)
29	30		149
37	60		140
34	90 ↓	↑	133
12	100		129
+27	120	then →	127

Elementary Example #2

Count quickly, as follows:

34	30		100 (the answer)
26	50		96
17	60		90
5	↓	↑	83
12	70		78
+6		then →	76

Brain Builder #1

Count quickly, as follows:

73	70		223 (the answer)
14	80		220
66	140		216
38	170 ↓	↑	210
7			202
+25	190	then →	195

Brain Builder #2

Count quickly, as follows:

11	10		263 (the answer)
75	80		262
34	110		257
68	170 ↓	↑	253
49	210		245
+26	230	then →	236

Number-Power Note: As previously stated, being able to mentally add a column of numbers is one of the most useful number-power tools you can acquire. Choose either Variation 1 (Trick 35) or Variation 2 (Trick 36), whichever is more to your liking. This method will also work, of course, when the addends are in the hundreds, or even in the thousands; however, far more concentration is needed to pull off the execution.

Elementary Exercises

Remember to perform the addition in a giant "U."

1.		4.		7.		10.	
	19		24		31		30
	33		9		10		5
	7		18		1		18
	41		49		43		17
	1		5		29		21
	+27		+35		+ 8		+45

2.		5.		8.		11.	
	3		7		31		42
	47		17		17		25
	30		46		5		3
	8		10		28		14
	13		2		2		33
	+22		+23		+56		+ 7

3.		6.		9.		12.	
	46		29		24		37
	28		52		27		2
	6		33		6		26
	32		10		13		15
	17		6		38		41
	+ 2		+45		+ 9		+28

13.	14.	15.	16.
2	23	26	11
35	45	33	20
9	30	7	5
19	5	19	36
42	1	47	27
+23	+18	+ 3	+ 6

Brain Builders

1.	4.	7.	10.
57	41	47	25
7	25	85	17
16	4	6	26
21	12	14	4
68	37	26	59
+34	+84	+31	+7

2.	5.	8.
21	42	57
49	37	35
36	28	9
50	10	19
6	1	22
+13	+56	+70

3.	6.	9.
42	38	60
5	2	19
14	63	5
51	19	51
83	26	13
+39	+41	+44

(See solutions on page 210)

Trick 37: Rapidly Add with the "Cross-Out" Technique

Strategy: This technique is especially useful when adding long columns of numbers, where accumulations can become uncomfortably large. To execute this trick, add in the usual manner, but **every time you reach or exceed ten**, lightly **cross out that digit** and **resume with just the ones digit**. For example, if you were to encounter the problem 8 + 7, you would focus on just the ones digit of the sum (5), and lightly cross out the 7 to indicate that 10 has been reached. The number of digits crossed out will then need to be carried over to the next column. Let's check out the following examples.

Elementary Example #1

Add:	→	Think:
4	→	4
~~8~~	→	2 (cross out 8)
3	→	5
~~7~~	→	2 (cross out 7)
0	→	skip
6	→	8
~~4~~	→	2 (cross out 4)
1	→	3
+ ~~7~~	→	0 (cross out 7)
4 0		

(Four ones digits crossed out translates to 4 carried to tens column.)

Elementary Example #2

Think:	←	Add:	→	Think:
		5		
2	←	~~7~~7	→	7
3	←	1~~6~~	→	3
7	←	42	→	5
5	←	~~8~~~~9~~	→	4
1	←	~~6~~5	→	9
4	←	3~~6~~	→	5
9	←	5~~5~~	→	0
6	←	~~7~~9	→	9
2	←	+ ~~6~~~~8~~	→	7
		527		

(Five tens digits crossed out translates to 5 carried to hundreds column.)

Brain Builder #1

Add:	⟶	Think:
4 4		
3~~8~~8		7 2 8
~~5~~1~~7~~		2 3 5
~~8~~0~~6~~		0 – 1
123		1 5 4
2~~5~~4		3 0 8
641		9 4 9
~~9~~~~9~~0		8 3 –
~~4~~3~~9~~		2 6 8
+ 5~~7~~~~5~~		7 3 3
4,733		

Brain Builder #2

Add:	⟶	Think:
4 5		
305		7 5 5
~~4~~4~~8~~		1 9 3
2~~1~~6		3 0 9
~~7~~8~~9~~		0 8 8
1~~2~~~~8~~		1 0 6
290		3 9 –
~~8~~~~8~~~~9~~		1 7 5
4~~5~~4		5 2 9
+ ~~5~~6~~7~~		0 8 6
4,086		

Number-Power Note: Another way to apply this technique is to use dots next to the digits, rather than lightly crossing them out. Use whichever way is faster and easier for you.

Elementary Exercises

Warning: counting any higher than nine is simply not allowed here.

1.	4.	7.	10.
8	95	88	78
7	68	43	26
2	88	96	87
1	36	14	91
6	51	50	16
2	20	73	35
3	77	28	65
9	49	69	84
+5	+15	+72	+23

2.	5.	8.	11.
74	46	51	55
18	94	25	37
85	63	67	98
53	27	82	56
90	11	19	78
24	55	80	41
66	99	54	93
71	70	32	48
+47	+34	+61	+95

3.	6.	9.	12.
52	75	23	21
79	21	75	59
29	64	45	27
42	76	30	89
60	83	12	35
13	17	89	40
97	38	52	72
44	86	31	37
+33	+62	+58	+66

13.	14.	15.	16.
96	72	37	53
53	58	55	32
93	97	76	95
16	60	14	18
44	12	83	68
78	41	20	34
49	86	99	29
23	38	46	71
+80	+91	+62	+48

Brain Builders

1.	4.	7.	9.
319	665	253	722
566	833	726	326
873	244	964	968
927	197	199	183
104	713	320	471
653	406	438	438
498	924	514	567
272	392	667	846
+795	+597	+855	+612

2.	5.	8.	10.
638	590	535	729
416	127	174	132
123	996	717	548
945	382	966	456
808	437	324	635
391	505	409	261
728	841	753	491
576	269	858	878
+252	+634	+681	+762

3.	6.
784	608
451	559
394	284
177	339
902	414
815	987
263	165
588	789
+686	+837

(See solutions on page 211)

Number Potpourri #5

When the odds of an event are placed at 2 to 1, it means the probability is one chance in how many? (See answer on page 225)

Trick 38: Rapidly Add Columns of Numbers in Sections

Strategy: When you have an unusually long column of numbers, there is a high probability that you will lose count or make a mistake. Accordingly, if you **divide the column into sections**, the computation becomes far more manageable. The number of sections to create is entirely up to you. Let's see how this handy trick works.

Elementary Example #1

```
 48
 17
 99
 30   →     194   (subtotal)
----
 82
 27
 66
 13   →     188   (subtotal)
----
 58
 33
 96
+75   →    +262   (subtotal)
----       ----
            644   (answer)
```

Elementary Example #2

$$
\begin{array}{rcrl}
83 & & & \\
44 & & & \\
16 & & & \\
59 & & & \\
30 & & & \\
\underline{63} & \rightarrow & 295 & \text{(subtotal)} \\
78 & & & \\
24 & & & \\
97 & & & \\
46 & & & \\
71 & & & \\
\underline{+36} & \rightarrow & \underline{+352} & \text{(subtotal)} \\
 & & 647 & \text{(answer)}
\end{array}
$$

Brain Builder #1

$$
\begin{array}{rcrl}
726 & & & \\
305 & & & \\
419 & & & \\
585 & & & \\
\underline{295} & \rightarrow & 2,330 & \text{(subtotal)} \\
763 & & & \\
486 & & & \\
274 & & & \\
832 & & & \\
\underline{748} & \rightarrow & 3,103 & \text{(subtotal)} \\
216 & & & \\
789 & & & \\
128 & & & \\
290 & & & \\
\underline{+889} & \rightarrow & \underline{+2,312} & \text{(subtotal)} \\
 & & 7,745 & \text{(answer)}
\end{array}
$$

Brain Builder #2

4,045			
6,749			
3,218			
6,660	→	20,672	(subtotal)
1,938			
9,815			
5,030			
2,465			
7,978	→	27,226	(subtotal)
8,641			
9,277			
3,429			
6,758			
+2,065	→	+30,170	(subtotal)
		78,068	(answer)

Number Power Note: This trick provides an easy mechanism for spotting mistakes, because you can check each shorter calculation until you discover where the error has been made.

Elementary Exercises

Try varying the number of sections you create for these.

1.	2.	3.	4.
19	38	84	65
66	16	51	33
73	23	94	44
27	45	77	97
24	88	62	13
53	91	15	76
98	28	63	24
72	76	88	92
95	52	86	97
+61	+67	+22	+81

5.
90
27
96
82
37
55
41
69
34
+18

6.
28
59
84
39
14
87
65
89
37
+46

7.
53
26
64
99
20
38
14
67
55
+82

8.
35
74
17
66
24
59
53
58
81
+92

9.
22
26
68
83
71
38
67
46
12
+59

10.
29
32
48
56
35
61
91
78
62
+17

11.
95
74
39
40
29
64
18
22
57
+81

12.
28
43
87
99
55
11
72
36
49
+64

13.
36
85
13
52
98
21
63
54
75
+49

14.
56
18
23
94
87
44
61
93
53
+35

15.
47
91
56
84
12
39
26
90
25
+77

16.
12
95
76
34
83
25
59
40
66
+81

Brain Builders

1.	2.	3.	4.	5.	6.	7.	8.	9.	10.
608	593	405	750	418	775	605	628	4,836	9,893
319	638	784	665	590	608	253	535	7,722	2,729
566	416	451	833	127	559	726	174	1,326	6,132
873	123	394	244	996	284	964	717	3,968	7,548
927	945	177	197	382	339	199	966	9,183	1,456
104	808	902	713	437	414	320	324	5,471	3,635
653	391	815	406	505	987	438	409	6,438	8,261
498	728	263	924	841	165	514	753	8,567	4,491
272	576	588	392	269	789	667	858	2,846	5,878
+795	+252	+686	+597	+634	+837	+855	+681	+5,612	+2,762

(See solutions on page 211)

Trick 39: Rapidly Add a Few Numbers

Strategy: When adding just a few numbers (preferably no more than four), it is normally fastest to **begin with the largest number** and **end with the smallest**. The theory is that it is easier to add a small number to a large one rather than the other way around. Be sure you have mastered either Trick 35 or 36 ("Rapidly Add a Column of Numbers") before proceeding with this trick. You will realize its value after you've looked over the following examples.

Elementary Example #1

3 + 64 + 38 + 5

Rearrange in your mind as: 64 + 38 + 5 + 3.

Applying Trick 35, think: 64, 74, 84, 94, 102, 107, 110 (the answer).

(*Note:* Obviously, shortcuts could have been taken in applying Trick 35. In fact, it could have been solved as: 64, 102, 110.)

Elementary Example #2

9 + 27 + 6 + 56

Rearrange in your mind as: 56 + 27 + 9 + 6.

Applying Trick 35, think: 56, 66, 76, 83, 92, 98 (the answer).

Brain Builder #1

19 + 144 + 5 + 33

Rearrange in your mind as: 144 + 33 + 19 + 5.

Applying Trick 35, think: 144, 177, 187, 196, 201 (the answer).

Brain Builder #2

71 + 45 + 124 + 12

Rearrange in your mind as: 124 + 71 + 45 + 12.

Applying Trick 35, think: 124, 195, 240, 252 (the answer).

Number-Power Note: This trick is especially useful when used with certain other tricks. For example, Trick 25 entails cross-multiplication and mental addition at a certain point. If you first perform the multiplication that produces the higher product, and add to it the smaller product, the overall calculation will go a lot more smoothly, with less chance of error. Finally, try performing the above calculations, but going from smallest to largest. You will then appreciate the power of this technique.

Elementary Exercises

Remember to go from large to small when working through these exercises.

1. $17 + 54 + 35 + 8 =$
2. $22 + 5 + 61 + 39 =$
3. $6 + 21 + 67 + 33 =$
4. $72 + 9 + 48 + 13 =$
5. $24 + 55 + 8 + 19 =$
6. $7 + 38 + 26 + 71 =$
7. $17 + 3 + 42 + 66 =$
8. $54 + 6 + 27 + 48 =$
9. $82 + 25 + 30 + 7 =$
10. $41 + 3 + 18 + 79 =$
11. $29 + 9 + 73 + 15 =$
12. $11 + 64 + 5 + 49 =$
13. $6 + 83 + 17 + 47 =$
14. $36 + 15 + 91 + 4 =$
15. $31 + 56 + 9 + 62 =$
16. $33 + 8 + 12 + 84 =$

Brain Builders

1. $47 + 6 + 139 + 22 =$
2. $31 + 128 + 7 + 63 =$
3. $2 + 117 + 55 + 26 =$
4. $164 + 37 + 8 + 11 =$
5. $44 + 15 + 147 + 63 =$
6. $78 + 24 + 131 + 9 =$
7. $7 + 175 + 60 + 34 =$
8. $29 + 106 + 33 + 65 =$
9. $16 + 72 + 119 + 47 =$
10. $193 + 12 + 68 + 4 =$

(See solutions on page 211)

Mathematical Curiosity #6

$$\begin{aligned} 34^2 &= 1156 \\ 334^2 &= 111556 \\ 3334^2 &= 11115556 \\ 33334^2 &= 1111155556 \end{aligned}$$

etc.

$$\begin{aligned} 67^2 &= 4489 \\ 667^2 &= 444889 \\ 6667^2 &= 44448889 \\ 66667^2 &= 4444488889 \end{aligned}$$

etc.

Trick 40: Rapidly Add 1 + 2 + 3, etc.

Strategy: This trick is especially useful in solving certain finance and accounting problems. To add $1 + 2 + 3 + \ldots + n$, multiply n by $(n + 1)$, and then divide by 2. For example, to add the numbers 1 through 8, calculate $(8 \times 9) \div 2 = 36$. Here are some more examples.

Elementary Example #1

$1 + 2 + 3 + \cdots + 7$

Apply the formula: $(7 \times 8) \div 2 = 28$ (the answer).

Elementary Example #2

$1 + 2 + 3 + \cdots + 10$

Apply the formula: $(10 \times 11) \div 2 = 55$ (the answer).

(Did you notice that it's fastest to perform the above calculation by dividing the 10 by 2, and then multiplying the resulting 5 by 11?)

Brain Builder #1

$1 + 2 + 3 + \cdots + 40$

Apply the formula: $(40 \times 41) \div 2 = 820$ (the answer).

(Reminder: In the above calculation, divide the 40 by 2, and multiply the resulting 20 by 41.)

Brain Builder #2

$1 + 2 + 3 + \cdots + 100$

Apply the formula: $(100 \times 101) \div 2 = 5,050$.

(Did you divide the 100 by 2, and then apply the "101 trick"?)

Number-Power Note: See the Mathematical Curiosity on page 149 for a variation on this technique.

Elementary Exercises

Just apply the formula, and you're home free.

1. $1 + 2 + 3 + 4 + 5 =$
2. $1 + 2 + 3 + 4 + 5 + 6 =$
3. $1 + 2 + 3 + \ldots + 8 =$
4. $1 + 2 + 3 + \ldots + 15 =$
5. $1 + 2 + 3 + \ldots + 12 =$
6. $1 + 2 + 3 + \ldots + 20 =$
7. $1 + 2 + 3 + \ldots + 17 =$
8. $1 + 2 + 3 + \ldots + 14 =$
9. $1 + 2 + 3 + \ldots + 25 =$
10. $1 + 2 + 3 + \ldots + 22 =$
11. $1 + 2 + 3 + \ldots + 18 =$
12. $1 + 2 + 3 + \ldots + 27 =$
13. $1 + 2 + 3 + \ldots + 21 =$
14. $1 + 2 + 3 + \ldots + 9 =$
15. $1 + 2 + 3 + \ldots + 30 =$
16. $1 + 2 + 3 + \ldots + 11 =$

Brain Builders

1. $1 + 2 + 3 + \ldots + 50 =$
2. $1 + 2 + 3 + \ldots + 35 =$
3. $1 + 2 + 3 + \ldots + 77 =$
4. $1 + 2 + 3 + \ldots + 90 =$
5. $1 + 2 + 3 + \ldots + 52 =$
6. $1 + 2 + 3 + \ldots + 81 =$
7. $1 + 2 + 3 + \ldots + 70 =$
8. $1 + 2 + 3 + \ldots + 58 =$
9. $1 + 2 + 3 + \ldots + 94 =$
10. $1 + 2 + 3 + \ldots + 63 =$

(See solutions on page 212)

Trick 41: Rapidly Subtract in Two Steps

Strategy: Tricks 29 and 30 recommend that you subtract by means of adding. However, when the calculation is presented in horizontal or verbal form, or if you find subtracting by adding just too difficult, try subtracting in two steps. That is, first subtract the tens digits, and then the ones. For example, to subtract 31 from 80, ignore the 1, and think, "80 − 30 = 50." Then think, "50 − 1 = 49" (the answer). In effect, you are subtracting from left to right when you use this trick. Let's look at a few more examples.

Elementary Example #1

70 − 42 =

Step 1. Ignore the 2, and think, "70 − 40 = 30."

Step 2. Subtract the 2: 30 − 2 = 28 (the answer).

Elementary Example #2

120 − 73 =

Step 1. Ignore the 3, and think, "120 − 70 = 50."

Step 2. Subtract the 3: 50 − 3 = 47 (the answer).

Brain Builder #1

93 − 57 =

Step 1. Ignore the 7, and think, "93 − 50 = 43."

Step 2. Subtract the 7: 43 − 7 = 36 (the answer).

Brain Builder #2

137 − 78 =

Step 1. Ignore the 8, and think, "137 − 70 = 67."

Step 2. Subtract the 8: 67 − 8 = 59 (the answer).

Number-Power Note: Brain Builder Examples 1 and 2 might be performed more easily if you initially ignore the ones digit of the minuend as well. For example, in Brain Builder#2, first think, "130 − 70 = 60." Then think, "67 − 8 = 59" (the answer). Or in Brain Builder#1, first think, "90 − 50 = 40." Then think, "43 − 7 = 36" (the answer).

Elementary Exercises

Remember to subtract the tens digits, and then the ones, when doing these exercises.

1. 90 − 66 =
2. 60 − 27 =
3. 40 − 18 =
4. 70 − 29 =
5. 100 − 34 =
6. 110 − 65 =
7. 120 − 83 =
8. 140 − 62 =
9. 80 − 52 =
10. 50 − 21 =
11. 100 − 27 =
12. 90 − 53 =
13. 70 − 36 =
14. 150 − 78 =
15. 130 − 69 =
16. 200 − 65 =

Brain Builders

1. 62 − 27 =
2. 87 − 39 =
3. 74 − 18 =
4. 93 − 44 =
5. 51 − 27 =
6. 108 − 35 =
7. 121 − 39 =
8. 136 − 89 =
9. 174 − 128 =
10. 300 − 237 =

(See solutions on page 212)

Number Potpourri #6

How carefully do you pay attention to numbers? We're going to put you on the spot.

What is the approximate population of the United States?

a. 32 million people
b. 87 million people
c. 160 million people
d. 250 million people

What is the approximate population of the world?

a. 822 million people
b. 5.4 billion people
c. 13.7 billion people
d. 28.1 billion people

(See answers on page 225)

Trick 42: Rapidly Check Addition and Subtraction

Strategy: You should review Trick 14 ("Rapidly Check Multiplication and Division") before proceeding with this trick. Trick 42 is applied in the same manner, except that the digit sums of the addends are **added**, not multiplied, and compared with the digit sum of the answer. Remember that when a calculation has been performed correctly, this "casting out nines" method (as it is usually called) will indicate as such. However, when the wrong answer has been obtained, this method will *probably*, but not definitely, uncover the error. The following examples will help clarify this terrific technique.

Elementary Example #1

22 + 14 = 36

22	(digit sum = 4)
+14	(digit-sum = 5)
____	(add: 4 + 5 = 9)
36	(digit sum = 9)

Because the third and fourth digit sums agree at 9, the answer is probably correct.

Elementary Example #2

86 + 75 = 151

86	(digit sum = 14; digit sum of 14 = 5)
+75	(digit sum = 12; digit sum of 12 = 3)
____	(add: 5 + 3 = 8)
151	(digit sum = 7)

Because the third and fourth digit sums of 8 and 7 do not agree, the answer is definitely incorrect.

Brain Builder #1

418 + 397 + 65 + 224 = 1,104

418	(digit sum = 13; digit sum of 13 = 4)
397	(digit sum = 19; digit sum of 19 = 10; digit sum of 10 = 1)
65	(digit sum = 11; digit sum of 11 = 2)
+224	(digit sum = 8)
_____	(add: 4 + 1 + 2 + 8 = 15; digit sum of 15 = 6)
1,104	(digit sum = 6)

Because the last two digit sums agree at 6, the answer is probably correct.

Number-Power Note: This technique has been illustrated for addition only. However, because subtraction is the inverse of addition, simply convert the subtraction into an addition, and apply the trick as already explained above. For example, to test 75 − 37 = 38, look at it as 38 + 37 = 75, and go from there.

Elementary Exercises

Check the calculations below for accuracy, using the "casting out nines" method. For each one, indicate "probably correct" or "definitely incorrect." It might help to reread the Number-Power Note to see how to check subtraction problems for accuracy.

1. $37 + 16 = 53$
2. $27 + 85 = 122$
3. $66 + 71 = 127$
4. $42 + 92 = 134$
5. $57 + 15 = 72$
6. $32 + 88 = 110$
7. $94 + 61 = 155$
8. $44 + 79 = 123$
9. $122 - 56 = 66$
10. $178 - 89 = 99$
11. $152 - 84 = 68$
12. $177 - 96 = 81$
13. $140 - 75 = 75$
14. $114 - 53 = 61$
15. $156 - 92 = 64$
16. $175 - 88 = 97$

Brain Builders

1. $719 + 422 + 905 + 72 = 2,018$
2. $187 + 672 + 80 + 425 = 1,364$
3. $48 + 665 + 194 + 373 = 1,280$
4. $948 + 205 + 167 + 720 = 2,140$
5. $281 + 777 + 546 + 15 = 1,619$
6. $327 + 801 + 213 + 946 = 2,288$
7. $158 + 764 + 99 + 432 = 1,453$
8. $444 + 398 + 206 + 50 = 998$
9. $526,897 - 378,165 = 148,732$
10. $83,107 - 44,659 = 38,548$
11. $99,236 - 56,742 = 42,494$
12. $840,653 - 598,771 = 241,882$
13. $5,391,403 - 2,648,772 = 2,752,631$
14. $11,724,396 - 9,801,557 = 1,922,839$

(See solutions on page 212)

Week 3 Quick Quiz

Let's see how many tricks from Week 3 you can remember and apply by taking this brief test. There's no time limit, but try to work through these items as rapidly as possible. Before you begin, glance at the computations and try to identify the trick that you could use. When you flip ahead to the solutions, you will see which trick was intended.

Elementary Problems

1. 59 + 73 =
2. 123 − 85 =
3. Use the "cross-out" technique:

   86
   62
   17
   38
   76
   +21

4. Add in two sections:

   92
   81
   53
   24
   77
   +69

5. 74 − 38 =
6. 1 + 2 + 3 + . . . + 20 =
7. 44 + 78 =
8. 4+7+1+8+1+3+5+9 =
9. 16 + 83 + 41 + 5 =
10. Using the "casting out nines" method, indicate whether this calculation is probably correct or definitely incorrect:

    32 + 89 = 131

11. 113 − 75 =
12. 130 − 62 =
13. 66 − 39 =
14. 87 − 35 =
15. 12
    33
    25
    7
    49
    +4

16. Add these without carrying:

    86
    69
    77
    35
    +81

Brain Builders

1. $135 - 77 =$
2. $176 + 69 =$
3. $7+9+3+1+7+8+5 =$
4. $1 + 2 + 3 + \ldots + 101 =$
5. $1,200 - 875 =$
6. $140 - 87 =$
7.
$$\begin{array}{r} 55 \\ 29 \\ 3 \\ 47 \\ 18 \\ +6 \\ \hline \end{array}$$
8.
$$\begin{array}{r} 40 \\ 17 \\ 4 \\ 36 \\ 29 \\ +8 \\ \hline \end{array}$$
9. $36 + 79 + 141 + 8 =$
10. $460 - 379 =$
11. Add in two sections:
$$\begin{array}{r} 4,971 \\ 862 \\ 307 \\ 3,553 \\ 2,499 \\ +616 \\ \hline \end{array}$$
12. Using the "casting out nines" method, indicate whether this calculation is probably correct or definitely incorrect:
$98,425 - 41,708 = 56,717$
13. Add these without carrying:
$$\begin{array}{r} 854 \\ 309 \\ 667 \\ 491 \\ 526 \\ +184 \\ \hline \end{array}$$
14. Use the "cross-out" technique:
$$\begin{array}{r} 722 \\ 418 \\ 339 \\ 245 \\ 866 \\ 192 \\ 508 \\ +771 \\ \hline \end{array}$$

(See solutions on page 222)

Week 4 Conversion Tricks and Estimation

What is a conversion trick? A term invented for this book, a conversion trick is one in which a calculation is presented in one form (decimal, fraction, or percentage), but solved using one of the other forms. Tricks 43 through 46 present the calculation in *decimal* form (the form most commonly seen in everyday life), but will make use of *fractions* to complete the calculation. You should first review the "Table of Equivalencies" on page 11 before proceeding to Trick 43.

Trick 43: Rapidly Multiply By 75 (or 0.75, 7.5, 750, etc.)

Strategy: Let's take a look at the first of four conversion tricks to be presented. To multiply a number by 75, **multiply the number by $\frac{3}{4}$**, and affix or insert any necessary zeroes or decimal point. The easiest way to multiply by $\frac{3}{4}$ is to multiply by $\frac{1}{2}$ (divide by 2), and then add half again. For example, to obtain $\frac{3}{4}$ of 28, take half of 28 (14), and add half again (7), to obtain the answer 21. The following examples should make this technique crystal clear.

Elementary Example #1
36 × 75

Step 1. Multiply: $36 \times \frac{3}{4} = 27$ (intermediary product).

Step 2. Apply T of R: A quick estimate puts the answer in the 2,000s.

Step 3. Affix two zeroes to the intermediary product, producing the answer 2,700.

Thought Process Summary

$$\begin{array}{r} 36 \\ \underline{\times 75} \end{array} \rightarrow 36 \times \tfrac{3}{4} = 27 \rightarrow 2{,}700$$

Elementary Example #2

24 × 75

Step 1. Multiply: $24 \times \frac{3}{4} = 18$ (intermediary product).

Step 2. Apply T of R: A quick estimate puts the answer in the 1,000s.

Step 3. Affix two zeroes to the intermediary product, producing the answer 1,800.

Thought Process Summary

$$\begin{array}{r} 24 \\ \underline{\times 75} \end{array} \rightarrow 24 \times \tfrac{3}{4} = 18 \rightarrow 1{,}800$$

Brain Builder #1

640 × 0.75

Step 1. Disregard the zero and decimal point and think, "64 × 75."

Step 2. Multiply: $64 \times \frac{3}{4} = 48$ (intermediary product).

Step 3. Apply T of R: A quick estimate puts the answer in the 400s.

Step 4. Affix one zero to the intermediary product, producing the answer 480.

Thought Process Summary

$$\begin{array}{r} 640 \\ \underline{\times 0.75} \end{array} \rightarrow \begin{array}{r} 64 \\ \underline{\times 75} \end{array} \rightarrow 64 \times \tfrac{3}{4} = 48 \rightarrow 480$$

Brain Builder #2

8.8 × 7.5

Step 1. Disregard the decimal points, and think, "88 × 75."

Step 2. Multiply: $88 \times \frac{3}{4} = 66$ (intermediary product).

Step 3. Apply T of R: A quick estimate puts the answer in the 60s. The intermediary product, 66, is also the correct answer.

Thought Process Summary

$$\begin{array}{r} 8.8 \\ \times 7.5 \\ \hline \end{array} \rightarrow \begin{array}{r} 88 \\ \times 75 \\ \hline \end{array} \rightarrow 88 \times \frac{3}{4} = 66$$

Number-Power Note: It is essential that you remember the trick to multiplying by $\frac{3}{4}$ described above—take half, and then add half again. You should spend some time practicing this trick alone as you learn Trick 43.

Elementary Exercises

Remember to multiply by $\frac{3}{4}$ when doing these exercises.

1. $28 \times 75 =$
2. $32 \times 75 =$
3. $44 \times 75 =$
4. $16 \times 75 =$
5. $75 \times 12 =$
6. $75 \times 52 =$
7. $75 \times 56 =$
8. $75 \times 8 =$
9. $48 \times 75 =$
10. $84 \times 75 =$

Brain Builders

1. $5.6 \times 750 =$
2. $320 \times 75 =$
3. $160 \times 7.5 =$
4. $8.4 \times 7.5 =$
5. $75 \times 22 =$
6. $75 \times 26 =$
7. $75 \times 30 =$
8. $75 \times 18 =$
9. $140 \times 7.5 =$
10. $5.2 \times 7.5 =$

(See solutions on page 213)

Number Potpourri # 7

Of 76, 152, and 246, which is (are) divisible by 8?

(See answer on page 225)

Trick 44: Rapidly Divide by 75 (or 0.75, 7.5, 750, etc.)

Strategy: You may recall that Trick 22 can be used to divide by 75. Here is an entirely different approach. To divide a number by 75, **multiply the number by $1\frac{1}{3}$**, and affix or insert any necessary zeroes or decimal point. The easiest way to multiply by $1\frac{1}{3}$ is to take $\frac{1}{3}$ of the number, and then add the number itself. For example, $1\frac{1}{3}$ of 18 equals 6 plus 18, or 24. Let's look at a few more examples.

Elementary Example #1

210 ÷ 75

Step 1. Disregard the zero and think, "21 ÷ 75."

Step 2. Multiply: $21 \times 1\frac{1}{3} = 28$ (intermediary product).

Step 3. Apply T of R: A quick estimate puts the answer between 2 and 3.

Step 4. Insert a decimal point within the intermediary product, producing the answer 2.8

Thought Process Summary

210 ÷ 75 → 21 ÷ 75 → $21 \times 1\frac{1}{3} = 28$ → 2.8

Elementary Example #2

90 ÷ 75

Step 1. Multiply: $90 \times 1\frac{1}{3} = 120$ (intermediary product).

Step 2. Apply T of R: A quick estimate puts the answer slightly above 1.

Step 3. Insert a decimal point within the intermediary product, producing the answer 1.2 (Note: The zero in the 90 could have been disregarded, but it didn't seem necessary to complete the calculation.)

Thought Process Summary

90 ÷ 75 → $90 \times 1\frac{1}{3} = 120$ → 1.2

Brain Builder #1

120 ÷ 7.5

Step 1. Disregard the zero and decimal point, and think, "12 ÷ 75."

Step 2. Multiply: $12 \times 1\frac{1}{3} = 16$ (intermediary product).

Step 3. Apply T of R: A quick estimate puts the answer between 10 and 20. The intermediary product, 16, is the answer.

Thought Process Summary

$120 \div 7.5 \rightarrow 12 \div 75 \rightarrow 12 \times 1\frac{1}{3} = 16$

Brain Builder #2

240 ÷ 750

Step 1. Disregard the zeroes and think, "24 ÷ 75."

Step 2. Multiply: $24 \times 1\frac{1}{3} = 32$ (intermediary product).

Step 3. Apply T of R: A quick estimate puts the answer between 0 and 1.

Step 4. Affix a decimal point to the left of the intermediary product, producing the answer 0.32.

Thought Process Summary

$240 \div 750 \rightarrow 24 \div 75 \rightarrow 24 \times 1\frac{1}{3} = 32 \rightarrow 0.32$

Number-Power Note: It is essential that you remember the trick to multiply by $1\frac{1}{3}$ described above—take $\frac{1}{3}$, and then add the number itself.

Elementary Exercises

When working through these exercises, remember to multiply by $1\frac{1}{3}$.

1. $360 \div 75 =$
2. $1,500 \div 75 =$
3. $2,700 \div 75 =$
4. $330 \div 75 =$
5. $420 \div 75 =$
6. $2,100 \div 75 =$
7. $1,800 \div 75 =$
8. $600 \div 75 =$
9. $660 \div 75 =$
10. $3,900 \div 75 =$
11. $4,500 \div 75 =$
12. $30 \div 75 =$
13. $510 \div 75 =$
14. $1,200 \div 75 =$
15. $9,900 \div 75 =$
16. $480 \div 75 =$

Brain Builders

1. $9,300 \div 75 =$
2. $720 \div 75 =$
3. $870 \div 75 =$
4. $6,900 \div 75 =$
5. $180 \div 7.5 =$
6. $450 \div 750 =$
7. $66 \div 75 =$
8. $5,400 \div 75 =$
9. $96 \div 75 =$
10. $6.3 \div 7.5 =$

(See solutions on page 213)

Trick 45: Rapidly Divide by 8 (or 0.8, 80, 800, etc.)

Strategy: We begin day 23 with another interesting conversion trick. To divide a number by 8, **multiply the number by $1\frac{1}{4}$**, and affix or insert any necessary zeroes or decimal point. The easiest way to multiply by $1\frac{1}{4}$ is to take $\frac{1}{4}$ of the number, and then add the number itself. For example, $1\frac{1}{4}$ of 24 equals 6 plus 24, or 30. Don't forget to make use of Trick 4 to determine $\frac{1}{4}$ of a number. The following examples illustrate this very practical trick.

Elementary Example #1

20 ÷ 8

Step 1. Multiply: $20 \times 1\frac{1}{4} = 25$ (intermediary product).

Step 2. Apply T of R: A quick estimate puts the answer between 2 and 3.

Step 3. Insert a decimal point within the intermediary product, producing the answer 2.5 (Note: One could also reduce the above problem to 5 ÷ 2.)

Thought Process Summary

$20 \div 8 \rightarrow \quad 20 \times 1\frac{1}{4} = 25 \quad \rightarrow \quad 2.5$

Elementary Example #2

28 ÷ 8

Step 1. Multiply: $28 \times 1\frac{1}{4} = 35$ (intermediary product).

Step 2. Apply T of R: A quick estimate puts the answer between 3 and 4.

Step 3. Insert a decimal point within the intermediary product, producing the answer 3.5.

Thought Process Summary

$28 \div 8 \rightarrow 28 \times 1\frac{1}{4} = 35 \rightarrow 3.5$

Brain Builder #1

220 ÷ 8

Step 1. Disregard the zero and think, "22 ÷ 8."

Step 2. Multiply: $22 \times 1\frac{1}{4} = 27.5$ (intermediary product).

Step 3. Apply T of R: A quick estimate puts the answer between 20 and 30. The intermediary product, 27.5, is the answer.

Thought Process Summary

$220 \div 8 \rightarrow 22 \div 8 \rightarrow 22 \times 1\frac{1}{4} = 27.5$

Brain Builder #2

1.8 ÷ 0.8

Step 1. Disregard the decimal points, and think, "18 ÷ 8."

Step 2. Multiply: $18 \times 1\frac{1}{4} = 22.5$ (intermediary product).

Step 3. Apply T of R: A quick estimate puts the answer just above 2.

Step 4. Move the decimal point of the intermediary product one place to the left, producing the answer 2.25.

Thought Process Summary

$1.8 \div 0.8 \rightarrow 18 \div 8 \rightarrow 18 \times 1\frac{1}{4} = 22.5 \rightarrow 2.25$

Number-Power Note: The conversion factor ($1\frac{1}{4}$) is easy to remember because you may recall that 8 and 125 are reciprocals, and 1.25 is equivalent to $1\frac{1}{4}$. Another way to divide a number by 8, of course, is to take half of the number, take half again, and finally halve again.

Elementary Exercises

When working these exercises, remember to multiply by $1\frac{1}{4}$.

1. $36 \div 8 =$
2. $52 \div 8 =$
3. $44 \div 8 =$
4. $50 \div 8 =$
5. $60 \div 8 =$
6. $140 \div 8 =$
7. $68 \div 8 =$
8. $30 \div 8 =$
9. $84 \div 8 =$
10. $180 \div 8 =$
11. $100 \div 8 =$
12. $260 \div 8 =$
13. $120 \div 8 =$
14. $76 \div 8 =$
15. $92 \div 8 =$
16. $280 \div 8 =$

Brain Builders

1. $4{,}400 \div 80 =$
2. $340 \div 8 =$
3. $920 \div 800 =$
4. $2{,}000 \div 80 =$
5. $7.6 \div 8 =$
6. $500 \div 80 =$
7. $3{,}000 \div 800 =$
8. $104 \div 8 =$
9. $3.6 \div 0.8 =$
10. $1{,}200 \div 80 =$

(See solutions on page 213)

Mathematical Curiosity #7

$$\begin{aligned} 1+3 &= 4 \text{ (or } 2^2) \\ 1+3+5 &= 9 \text{ (or } 3^2) \\ 1+3+5+7 &= 16 \text{ (or } 4^2) \\ 1+3+5+7+9 &= 25 \text{ (or } 5^2) \\ & \textit{etc.} \end{aligned}$$

(Formula: To obtain the sum of successive odd numbers, simply count the number of numbers, and square it.)

Trick 46: Rapidly Divide by 15 (or 0.15, 1.5, 150, etc.)

Strategy: Here is the last, and perhaps the best, of the four conversion tricks presented. To divide a number by 15, **multiply the number by** $\frac{2}{3}$, and affix or insert any necessary zeroes or decimal point. The easiest way to multiply by $\frac{2}{3}$ is to take $\frac{1}{3}$ of the number, and then double it. For example, $\frac{2}{3}$ of 12 equals 4 doubled, or 8. Read on to see this trick in action.

Elementary Example #1

180 ÷ 15

Step 1. Disregard the zero and think, "18 ÷ 15."

Step 2. Multiply: $18 \times \frac{2}{3} = 12$ (intermediary product).

Step 3. Apply T of R: A quick estimate puts the answer between 10 and 20. The intermediary product, 12, is therefore the answer.

Thought Process Summary

$180 \div 15 \rightarrow 18 \div 15 \rightarrow 18 \times \frac{2}{3} = 12$

Elementary Example #2

240 ÷ 15

Step 1. Disregard the zero and think, "24 ÷ 15."

Step 2. Multiply: $24 \times \frac{2}{3} = 16$ (intermediary product).

Step 3. Apply T of R: A quick estimate puts the answer between 10 and 20. The intermediary product, 16, is therefore the answer.

Thought Process Summary

$240 \div 15 \rightarrow 24 \div 15 \rightarrow 24 \times \frac{2}{3} = 16$

Brain Builder #1

39 ÷ 15

Step 1. Multiply: $39 \times \frac{2}{3} = 26$ (intermediary product).

Step 2. Apply T of R: A quick estimate puts the answer between 2 and 3.

Step 3. Insert a decimal point within the intermediary product, producing the answer 2.6.

Thought Process Summary

$39 \div 15 \rightarrow 39 \times \frac{2}{3} = 26 \rightarrow 2.6$

Brain Builder #2

72 ÷ 1.5

Step 1. Disregard the decimal point, and think, "72 ÷ 15."

Step 2. Multiply: $72 \times \frac{2}{3} = 48$ (intermediary product).

Step 3. Apply T of R: A quick estimate puts the answer around 50. The intermediary product, 48, is therefore the answer.

Thought Process Summary

$$72 \div 1.5 \rightarrow 72 \div 15 \rightarrow 72 \times \frac{2}{3} = 48$$

Number-Power Note: This method is an alternative to Trick 22 "rapidly divide by 1.5."

Elementary Exercises

Remember to multiply by $\frac{2}{3}$ when working through these exercises.

1. 540 ÷ 15 =
2. 420 ÷ 15 =
3. 99 ÷ 15 =
4. 360 ÷ 15 =
5. 210 ÷ 15 =
6. 840 ÷ 15 =
7. 630 ÷ 15 =
8. 270 ÷ 15 =
9. 120 ÷ 15 =
10. 960 ÷ 15 =
11. 480 ÷ 15 =
12. 780 ÷ 15 =
13. 330 ÷ 15 =
14. 690 ÷ 15 =
15. 930 ÷ 15 =
16. 570 ÷ 15 =

Brain Builders

1. 81 ÷ 1.5 =
2. 660 ÷ 150 =
3. 240 ÷ 150 =
4. 8.7 ÷ 1.5 =
5. 5,100 ÷ 150 =
6. 9 ÷ 15 =
7. 45 ÷ 1.5 =
8. 9,300 ÷ 1,500 =
9. 3.6 ÷ 1.5 =
10. 78 ÷ 150 =

(See solutions on page 214)

A word about estimation: The most valuable number-power skill of them all is rapid estimation. Every day we encounter situations where we must produce a "ballpark figure" on the spot.

One estimation technique employed throughout this book is "rounding." That is, when it is necessary to apply a "test of reasonableness" to an intermediary figure, rounding will usually help us determine how many zeroes to affix or where to insert the decimal point.

Another estimation technique involves what might be called "approximate reciprocals." You may recall that "reciprocals," as the term is used in this book, are two numbers which "multiply out" to 10, 100, or some other multiple of 10. Approximate reciprocals are two numbers which *almost* multiply out to 10, 100, or the like. For example, 33 and 3 produce 99 when multiplied, as do 9 and 11. Therefore, to approximate multiplication by 33, divide by 3, and so on.

You are about to learn ten estimation techniques using approximate reciprocals (Tricks 47–56). Many more could have been presented, but I have found these to be the ten most useful and most easily remembered ones. Further, every trick presented will produce an estimate that is within 2 percent of the true answer (in some cases, the estimate will be within $\frac{1}{2}$ of 1 percent of the true answer).

When you've completed the ten estimation techniques, there will be only four more tricks for you to learn. Hang in there, and keep up the good work!

Trick 47: Rapidly Estimate Multiplication by 33 or 34

Strategy: Let's begin Day 24 with the first of 10 estimation techniques. To estimate multiplication by 33 or 34, **divide by 3**, and affix or insert any necessary zeroes or decimal point. This technique will also work when multiplying by 0.33, 3.3, 330, 0.34, 3.4, 340, and so on. The following examples demonstrate that this is a very easy trick to apply.

Elementary Example #1

Estimate 24 × 33

Step 1. Divide: 24 ÷ 3 = 8 (intermediary estimate).

Step 2. Apply T of R: A quick estimate puts the answer in the upper hundreds.

Step 3. Affix two zeroes to the intermediary estimate, producing a final estimate of 800 (the true answer is 792).

Thought Process Summary

$$\begin{array}{r} 24 \\ \times 33 \\ \hline \end{array} \rightarrow 24 \div 3 = 8 \rightarrow 800$$

Elementary Example #2

Estimate 18 × 34

Step 1. Divide: 18 ÷ 3 = 6 (intermediary estimate).

Step 2. Apply T of R: A quick estimate puts the answer in the middle hundreds.

Step 3. Affix two zeroes to the intermediary estimate, producing a final estimate of 600 (the true answer is 612).

Thought Process Summary

$$\begin{array}{r} 18 \\ \times 34 \\ \hline \end{array} \rightarrow 18 \div 3 = 6 \rightarrow 600$$

Brain Builder #1

Estimate 420 × 3.3

Step 1. Disregard the zero and decimal point, and think, "42 × 33."

Step 2. Divide: 42 ÷ 3 = 14 (intermediary estimate).

Step 3. Apply T of R: A quick estimate puts the answer between 1,000 and 2,000.

Step 4. Affix two zeroes to the intermediary estimate, producing a final estimate of 1,400 (the true answer is 1,386).

Thought Process Summary

$$\begin{array}{r} 420 \\ \times 3.3 \\ \hline \end{array} \rightarrow \begin{array}{r} 42 \\ \times 33 \\ \hline \end{array} \rightarrow 42 \div 3 = 14 \rightarrow 1,400$$

Brain Builder #2

Estimate 5.1 × 34

Step 1. Disregard the decimal point and think, "51 × 34."

Step 2. Divide: 51 ÷ 3 = 17 (intermediary estimate).

Step 3. Apply T of R: A quick estimate puts the answer between 100 and 200.

Step 4. Affix one zero to the intermediary estimate, producing a final estimate of 170 (the true answer is 173.4).

Thought Process Summary

$$\begin{array}{r} 5.1 \\ \times 34 \\ \hline \end{array} \rightarrow \begin{array}{r} 51 \\ \times 34 \\ \hline \end{array} \rightarrow 51 \div 3 = 17 \rightarrow 170$$

Number-Power Note: When using this technique to multiply by 33, your estimate will be about 1 percent over the true answer. When multiplying by 34, your estimate will be about 2 percent under the true answer. In the unusual event that you multiply by $3\frac{1}{3}$, $33\frac{1}{3}$, or the like, this technique of dividing by 3 will produce an *exact* answer.

Elementary Exercises

For all estimation-trick exercises, the symbol ≈ means "approximately equals." Remember to divide by 3 when solving these problems.

1. 15 × 33 ≈
2. 66 × 33 ≈
3. 48 × 34 ≈
4. 96 × 34 ≈
5. 33 × 57 ≈
6. 33 × 21 ≈
7. 34 × 72 ≈
8. 34 × 84 ≈
9. 93 × 33 ≈
10. 45 × 33 ≈
11. 27 × 34 ≈
12. 54 × 34 ≈
13. 33 × 78 ≈
14. 33 × 33 ≈
15. 34 × 75 ≈
16. 34 × 69 ≈

Brain Builders

1. $87 \times 3.3 \approx$
2. $3.9 \times 33 \approx$
3. $1.2 \times 340 \approx$
4. $99 \times 3.4 \approx$
5. $330 \times 2.4 \approx$
6. $0.33 \times 630 \approx$
7. $3.4 \times 540 \approx$
8. $3.4 \times 8.1 \approx$
9. $60 \times 3.3 \approx$
10. $180 \times 3.3 \approx$

(See solutions on page 214)

Number Potpourri #8

Liquor that is "50-proof" is what percent alcohol?

(See answer on page 225)

Trick 48: Rapidly Estimate Division by 33 or 34

Strategy: You've probably already figured out how this trick is going to work. To estimate division by 33 or 34, **multiply by 3**, and affix or insert any necessary zeroes or decimal point. This technique will also work when dividing by 0.33, 3.3, 330, 0.34, 3.4, 340, and so on. Here are some examples using this excellent estimation technique.

Elementary Example #1

Estimate 120 ÷ 33

Step 1. Disregard the zero and think, "12 ÷ 33."

Step 2. Multiply: $12 \times 3 = 36$ (intermediary estimate).

Step 3. Apply T of R: A quick estimate puts the answer between 3 and 4.

Step 4. Insert a decimal point within the intermediary estimate, producing a final estimate of 3.6 (the true answer is about 3.64).

Thought Process Summary

$$120 \div 33 \rightarrow 12 \div 33 \rightarrow \begin{array}{r} 12 \\ \times 3 \\ \hline 36 \end{array} \rightarrow 3.6$$

Elementary Example #2

Estimate 260 ÷ 34

Step 1. Disregard the zero and think, "26 ÷ 34."

Step 2. Multiply: 26 × 3 = 78 (intermediary estimate).

Step 3. Apply T of R: A quick estimate puts the answer between 7 and 8.

Step 4. Insert a decimal point within the intermediary estimate, producing a final estimate of 7.8 (the true answer is about 7.65).

Thought Process Summary

$$260 \div 34 \rightarrow 26 \div 34 \rightarrow \begin{array}{r} 26 \\ \times 3 \\ \hline 78 \end{array} \rightarrow 7.8$$

Brain Builder #1

Estimate 190 ÷ 3.3

Step 1. Disregard the zero and decimal point and think, "19 ÷ 33."

Step 2. Multiply: 19 × 3 = 57 (intermediary estimate).

Step 3. Apply T of R: A quick estimate puts the answer between 50 and 60. The intermediary estimate of 57 is correct (the true answer is about 57.58).

Thought Process Summary

$$190 \div 3.3 \rightarrow 19 \div 33 \rightarrow \begin{array}{r} 19 \\ \times 3 \\ \hline 57 \end{array}$$

Brain Builder #2

Estimate 3,200 ÷ 340

Step 1. Disregard the zeroes and think, "32 ÷ 34."

Step 2. Multiply: 32 × 3 = 96 (intermediary estimate).

Step 3. Apply T of R: A quick estimate puts the answer just under 10.

Step 4. Insert a decimal point within the intermediary estimate, producing a final estimate of 9.6 (the true answer is about 9.41).

Thought Process Summary

$$3{,}200 \div 340 \quad \rightarrow \quad 32 \div 34 \quad \rightarrow \quad \begin{array}{r} 32 \\ \times 3 \\ \hline 96 \end{array} \quad \rightarrow \quad 9.6$$

Number-Power Note: When using this technique to divide by 33, your estimate will be 1 percent under the true answer. When dividing by 34, your estimate will be 2 percent over the true answer. In the unusual event that you divide by $3\frac{1}{3}$, $33\frac{1}{3}$, or the like, this technique of multiplying by 3 will produce an *exact* answer.

Elementary Exercises

When working these exercises, remember to multiply by 3. Once again, the symbol $\approx$ means "approximately equals."

1. $700 \div 33 \approx$
2. $110 \div 33 \approx$
3. $250 \div 34 \approx$
4. $800 \div 34 \approx$
5. $160 \div 33 \approx$
6. $230 \div 33 \approx$
7. $400 \div 34 \approx$
8. $140 \div 34 \approx$
9. $310 \div 33 \approx$
10. $500 \div 33 \approx$
11. $750 \div 34 \approx$
12. $180 \div 34 \approx$
13. $900 \div 33 \approx$
14. $220 \div 33 \approx$
15. $280 \div 34 \approx$
16. $600 \div 34 \approx$

Brain Builders

1. $2{,}000 \div 330 \approx$
2. $17 \div 3.3 \approx$
3. $430 \div 34 \approx$
4. $820 \div 34 \approx$
5. $360 \div 3.3 \approx$
6. $1{,}300 \div 330 \approx$
7. $2.7 \div 0.34 \approx$
8. $450 \div 34 \approx$
9. $2{,}400 \div 33 \approx$
10. $91 \div 3.3 \approx$

(See solutions on page 215)

Trick 49: Rapidly Estimate Multiplication by 49 or 51

Strategy: Today's estimation techniques use the same strategy as those in Tricks 5 and 6. To estimate multiplication by 49 or 51, **divide by 2**, and affix or insert any necessary zeroes or decimal point. This trick will also work when multiplying by 0.49, 4.9, 490, 0.51, 5.1, 510, and so on. The exercises below demonstrate how wonderfully this estimation technique works.

Elementary Example #1

Estimate 62 × 49

Step 1. Divide: $62 \div 2 = 31$ (intermediary estimate).

Step 2. Apply T of R: A quick estimate puts the answer at about 3,000.

Step 3. Affix two zeroes to the intermediary estimate, producing a final estimate of 3,100 (the true answer is 3,038).

Thought Process Summary

$$\begin{array}{r} 62 \\ \times 49 \\ \hline \end{array} \rightarrow 62 \div 2 = 31 \rightarrow 3,100$$

Elementary Example #2

Estimate 35 × 51

Step 1. Divide: $35 \div 2 = 17.5$ (intermediary estimate).

Step 2. Apply T of R: A quick estimate puts the answer in the middle thousands.

Step 3. Eliminate the decimal point from the intermediary estimate and affix one zero, producing a final estimate of 1,750 (the true answer is 1,785).

Thought Process Summary

$$\begin{array}{r} 35 \\ \times 51 \\ \hline \end{array} \rightarrow 35 \div 2 = 17.5 \rightarrow 1,750$$

Brain Builder #1

Estimate 4.4 × 4.9

Step 1. Disregard the decimal points and think, "44 × 49."

Step 2. Divide: 44 ÷ 2 = 22 (intermediary estimate).

Step 3. Apply T of R: A quick estimate puts the answer in the 20s. The intermediary estimate of 22 is therefore correct (the true answer is 21.56).

Thought Process Summary

$$\begin{array}{r} 4.4 \\ \times 4.9 \\ \hline \end{array} \rightarrow \begin{array}{r} 44 \\ \times 49 \\ \hline \end{array} \rightarrow 44 \div 2 = 22$$

Brain Builder #2

Estimate 270 × 0.51

Step 1. Disregard the zero and decimal point and think, "27 × 51."

Step 2. Divide: 27 ÷ 2 = 13.5 (intermediary estimate).

Step 3. Apply T of R: A quick estimate puts the answer at just over half of 270, or at about 135.

Step 4. Eliminate the decimal point from the intermediary estimate, producing a final estimate of 135 (the true answer is 137.7).

Thought Process Summary

$$\begin{array}{r} 270 \\ \times 0.51 \\ \hline \end{array} \rightarrow \begin{array}{r} 27 \\ \times 51 \\ \hline \end{array} \rightarrow 27 \div 2 = 13.5 \rightarrow 135$$

Number-Power Note: When using this technique to multiply by 49, your estimate will be about 2 percent *over* the true answer. When multiplying by 51, your estimate will be about 2 percent *under* the true answer. As you know from having learned Trick 5, using this technique to multiply by 50 will produce an *exact* answer.

Elementary Exercises

Remember to divide by 2 when doing these exercises.

1. $64 \times 49 \approx$
2. $22 \times 49 \approx$
3. $78 \times 51 \approx$
4. $56 \times 51 \approx$
5. $49 \times 29 \approx$
6. $49 \times 13 \approx$
7. $51 \times 45 \approx$
8. $51 \times 31 \approx$
9. $56 \times 49 \approx$
10. $70 \times 49 \approx$
11. $92 \times 51 \approx$
12. $86 \times 51 \approx$
13. $49 \times 53 \approx$
14. $49 \times 37 \approx$
15. $51 \times 98 \approx$
16. $51 \times 71 \approx$

Brain Builders

1. $16 \times 4.9 \approx$
2. $7.5 \times 49 \approx$
3. $2.4 \times 5.1 \approx$
4. $97 \times 51 \approx$
5. $490 \times 0.88 \approx$
6. $4.9 \times 110 \approx$
7. $5.1 \times 67 \approx$
8. $5.1 \times 8.3 \approx$
9. $180 \times 0.49 \approx$
10. $71 \times 4.9 \approx$

(See solutions on page 215)

Trick 50: Rapidly Estimate Division by 49 or 51

Strategy: Let's examine another estimation technique that's a real winner. To estimate division by 49 or 51, **multiply by 2**, and affix or insert any necessary zeroes or decimal point. This technique will also work when dividing by 0.49, 4.9, 490, 0.51, 5.1, 510, and so on. You probably don't need to see any examples to understand this trick, but here are some just to be sure.

Elementary Example #1

Estimate 130 ÷ 49

Step 1. Disregard the zero and think, "13 ÷ 49."

Step 2. Multiply: $13 \times 2 = 26$ (intermediary estimate).

Step 3. Apply T of R: A quick estimate puts the answer between 2 and 3.

Step 4. Insert a decimal point within the intermediary estimate, producing a final estimate of 2.6 (the true answer is about 2.65).

Thought Process Summary

$$130 \div 49 \rightarrow 13 \div 49 \rightarrow \begin{array}{r} 13 \\ \times 2 \\ \hline 26 \end{array} \rightarrow 2.6$$

Elementary Example #2

Estimate 85 ÷ 51

Step 1. Multiply: 85 × 2 = 170 (intermediary estimate).

Step 2. Apply T of R: A quick estimate puts the answer between 1 and 2.

Step 3. Insert a decimal point within the intermediary estimate, producing a final estimate of 1.7 (the true answer is about 1.67).

Thought Process Summary

$$85 \div 51 \rightarrow \begin{array}{r} 85 \\ \times 2 \\ \hline 170 \end{array} \rightarrow 1.7$$

Brain Builder #1

Estimate 66 ÷ 4.9

Step 1. Disregard the decimal point and think, "66 ÷ 49."

Step 2. Multiply: 66 × 2 = 132 (intermediary estimate).

Step 3. Apply T of R: A quick estimate puts the answer in the teens.

Step 4. Insert a decimal point within the intermediary estimate, producing a final estimate of 13.2 (the true answer is about 13.47).

Thought Process Summary

$$66 \div 4.9 \rightarrow 66 \div 49 \rightarrow \begin{array}{r} 66 \\ \times 2 \\ \hline 132 \end{array} \rightarrow 13.2$$

Brain Builder #2

Estimate 390 ÷ 510

Step 1. Disregard the zeroes and think, "39 ÷ 51."

Step 2. Multiply: 39 × 2 = 78 (intermediary estimate).

Step 3. Apply T of R: A quick estimate puts the answer between 0 and 1.

Step 4. Affix a decimal point to the left of the intermediary estimate, producing a final estimate of 0.78 (the true answer is about 0.765).

Thought Process Summary

$$390 \div 510 \rightarrow 39 \div 51 \rightarrow \begin{array}{r} 39 \\ \times 2 \\ \hline 78 \end{array} \rightarrow 0.78$$

Number-Power Note: When using this trick to divide by 49, your estimate will be 2 percent *under* the true answer. When dividing by 51, your estimate will be 2 percent *over* the true answer. As you know from having learned Trick 6, using the above technique to divide by 50 will produce an *exact* answer.

Elementary Exercises

When working through these exercises, remember to multiply by 2.

1. 240 ÷ 49 ≈
2. 91 ÷ 49 ≈
3. 88 ÷ 51 ≈
4. 450 ÷ 51 ≈
5. 570 ÷ 49 ≈
6. 320 ÷ 49 ≈
7. 79 ÷ 51 ≈
8. 610 ÷ 51 ≈
9. 180 ÷ 49 ≈
10. 94 ÷ 49 ≈
11. 730 ÷ 51 ≈
12. 210 ÷ 51 ≈
13. 300 ÷ 49 ≈
14. 360 ÷ 49 ≈
15. 770 ÷ 51 ≈
16. 640 ÷ 51 ≈

Brain Builders

1. 43 ÷ 4.9 ≈
2. 970 ÷ 490 ≈
3. 1,500 ÷ 51 ≈
4. 70 ÷ 5.1 ≈
5. 8.2 ÷ 0.49 ≈
6. 2,700 ÷ 49 ≈
7. 58 ÷ 5.1 ≈
8. 340 ÷ 51 ≈
9. 7,500 ÷ 49 ≈
10. 99 ÷ 4.9 ≈

(See solutions on page 216)

Parlor Trick #4: Rapidly Subtract the Squares of Any Two Numbers

Strategy: This parlor trick falls under the category of excellent rapid-math techniques that unfortunately have little practical use, except as parlor tricks. To subtract the squares of two numbers, add the two numbers together and then multiply the sum by the difference between the two numbers.

Demonstration Problem A

$8^2 - 7^2 = (8 + 7) \times (8 - 7) = 15 \times 1 = 15$

Demonstration Problem B

$37^2 - 26^2 = (37 + 26) \times (37 - 26) = 63 \times 11 = 693$

Give your brain a workout with these:

A. $7^2 - 6^2 =$

B. $10^2 - 7^2 =$

C. $12^2 - 8^2 =$

D. $15^2 - 10^2 =$

E. $6^2 - 4^2 =$

F. $18^2 - 12^2 =$

(See solutions on page 224)

Trick 51: Rapidly Estimate Multiplication by 66 or 67

Strategy: This trick is both an estimation technique and a conversion trick. To estimate multiplication by 66 or 67, **multiply by $\frac{2}{3}$**, and affix or insert any necessary zeroes or decimal point. This technique will also work when multiplying by 0.66, 6.6, 660, 0.67, 6.7, 670, and so on. Remember that the easiest way to multiply by $\frac{2}{3}$ is to take $\frac{1}{3}$ of the number and then double it. For example, $\frac{2}{3}$ of 15 equals 5 doubled, or 10. Let's see exactly how this trick works.

Elementary Example #1
Estimate 30 × 66

Step 1. Multiply: $30 \times \frac{2}{3} = 20$ (intermediary estimate).

Step 2. Apply T of R: A quick estimate puts the answer at about 2,000.

Step 3. Affix two zeroes to the intermediary estimate, producing a final estimate of 2,000 (the true answer is 1,980).

Thought Process Summary

$$\begin{array}{r} 30 \\ \underline{\times 66} \end{array} \rightarrow 30 \times \frac{2}{3} = 20 \rightarrow 2,000$$

Elementary Example #2
Estimate 42 × 67

Step 1. Multiply: $42 \times \frac{2}{3} = 28$ (intermediary estimate).

Step 2. Apply T of R: A quick estimate puts the answer near 3,000.

Step 3. Affix two zeroes to the intermediary estimate, producing a final estimate of 2,800 (the true answer is 2,814).

Thought Process Summary

$$\begin{array}{r} 42 \\ \times 67 \\ \hline \end{array} \rightarrow 42 \times \tfrac{2}{3} = 28 \rightarrow 2{,}800$$

Brain Builder #1

Estimate 5.4 × 6.6

Step 1. Disregard the decimal points and think, "54 × 66."

Step 2. Multiply: 54 × $\frac{2}{3}$ = 36 (intermediary estimate).

Step 3. Apply T of R: A quick estimate puts the answer in the 30s. The intermediary estimate of 36 is therefore correct (the true answer is 35.64).

Thought Process Summary

$$\begin{array}{r} 5.4 \\ \times 6.6 \\ \hline \end{array} \rightarrow \begin{array}{r} 54 \\ \times 66 \\ \hline \end{array} \rightarrow 54 \times \tfrac{2}{3} = 36$$

Brain Builder #2

Estimate 390 × 0.67

Step 1. Disregard the zero and decimal point, and think, "39 × 67."

Step 2. Multiply: 39 × $\frac{2}{3}$ = 26 (intermediary estimate).

Step 3. Apply T of R: A quick estimate puts the answer in the 200s.

Step 4. Affix one zero to the intermediary estimate, producing a final estimate of 260 (the true answer is 261.3).

Thought Process Summary

$$\begin{array}{r} 390 \\ \times 0.67 \\ \hline \end{array} \rightarrow \begin{array}{r} 39 \\ \times 67 \\ \hline \end{array} \rightarrow 39 \times \tfrac{2}{3} = 26 \rightarrow 260$$

Number-Power Note: When using this trick to multiply by 66, your estimate will be about 1 percent *over* the true answer. When multiplying by 67, your estimate will be about $\frac{1}{2}$ of 1 percent *under* the true answer. In the unusual event that you multiply by $6\frac{2}{3}$, $66\frac{2}{3}$, or the like, this technique of multiplying by $\frac{2}{3}$ will produce an *exact* answer.

Elementary Exercises

Remember to multiply by $\frac{2}{3}$ when doing these exercises.

1. $12 \times 66 \approx$
2. $45 \times 66 \approx$
3. $36 \times 67 \approx$
4. $21 \times 67 \approx$
5. $66 \times 90 \approx$
6. $66 \times 18 \approx$
7. $67 \times 75 \approx$
8. $67 \times 33 \approx$
9. $51 \times 66 \approx$
10. $93 \times 66 \approx$
11. $60 \times 67 \approx$
12. $24 \times 67 \approx$
13. $66 \times 27 \approx$
14. $66 \times 84 \approx$
15. $67 \times 63 \approx$
16. $67 \times 96 \approx$

Brain Builders

1. $4.8 \times 66 \approx$
2. $150 \times 6.6 \approx$
3. $8.7 \times 6.7 \approx$
4. $420 \times 0.67 \approx$
5. $6.6 \times 300 \approx$
6. $66 \times 81 \approx$
7. $67 \times 7.2 \approx$
8. $6.7 \times 57 \approx$
9. $7.8 \times 66 \approx$
10. $990 \times 0.66 \approx$

(See solutions on page 216)

Trick 52: Rapidly Estimate Division by 66 or 67

Strategy: This trick is the inverse of Trick 46 ("Rapidly Divide by 15"). To estimate division by 66 or 67, **multiply by 1.5 ($1\frac{1}{2}$)**, and affix or insert any necessary zeroes or decimal point. This trick will also work when dividing by 0.66, 6.6, 660, 0.67, 6.7, 670, and so on. Remember from Trick 19 that multiplying by 1.5 means taking the number and adding half again. Let's look at some examples of this ingenious estimation technique.

Elementary Example #1

Estimate 400 ÷ 66

Step 1. Disregard the zeroes and think, "4 ÷ 66."

Step 2. Multiply: $4 \times 1.5 = 6$ (intermediary estimate).

Step 3 Apply T of R: A quick estimate puts the answer at about 6. The intermediary estimate of 6 is therefore correct (the true answer is about 6.06).

Thought Process Summary

$$400 \div 66 \rightarrow 4 \div 66 \rightarrow \begin{array}{r} 4 \\ \times 1.5 \\ \hline 6 \end{array}$$

Elementary Example #2

Estimate 500 ÷ 67

Step 1. Disregard the zeroes and think, "5 ÷ 67."

Step 2. Multiply: 5 × 1.5 = 7.5 (intermediary estimate).

Step 3. Apply T of R: A quick estimate puts the answer between 7 and 8. The intermediary estimate of 7.5 is therefore correct (the true answer is about 7.46).

Thought Process Summary

$$500 \div 67 \rightarrow 5 \div 67 \rightarrow \begin{array}{r} 5 \\ \times 1.5 \\ \hline 7.5 \end{array}$$

Brain Builder #1

Estimate 800 ÷ 6.6

Step 1. Disregard the zeroes and decimal point and think, "8 ÷ 66."

Step 2. Multiply: 8 × 1.5 = 12 (intermediary estimate).

Step 3. Apply T of R: A quick estimate puts the answer slightly over 100.

Step 4. Affix one zero to the intermediary estimate, producing a final estimate of 120 (the true answer is about 121.21).

Thought Process Summary

$$800 \div 6.6 \rightarrow 8 \div 66 \rightarrow \begin{array}{r} 8 \\ \times 1.5 \\ \hline 12 \end{array} \rightarrow 120$$

Brain Builder #2

Estimate 44 ÷ 6.7

Step 1. Disregard the decimal point and think, "44 ÷ 67."

Step 2. Multiply: 44 × 1.5 = 66 (intermediary estimate).

Step 3. Apply T of R: A quick estimate puts the answer between 6 and 7.

Step 4. Insert a decimal point within the intermediary estimate, producing a final estimate of 6.6 (the true answer is about 6.57).

Thought Process Summary

$$44 \div 6.7 \rightarrow 44 \div 67 \rightarrow \begin{array}{r} 44 \\ \times 1.5 \\ \hline 66 \end{array} \rightarrow 6.6$$

Number-Power Note: When using this trick to divide by 66, your estimate will be 1 percent *under* the true answer. When dividing by 67, your estimate will be $\frac{1}{2}$ of 1 percent *over* the true answer. In the unusual event that you divide by $6\frac{2}{3}$, $66\frac{2}{3}$, or the like, this technique of multiplying by 1.5 will produce an *exact* answer.

Elementary Exercises

When doing these exercises, remember to multiply by 1.5 ($1\frac{1}{2}$).

1. 140 ÷ 66 ≈
2. 220 ÷ 66 ≈
3. 600 ÷ 67 ≈
4. 360 ÷ 67 ≈
5. 280 ÷ 66 ≈
6. 420 ÷ 66 ≈
7. 640 ÷ 67 ≈
8. 580 ÷ 67 ≈
9. 1,200 ÷ 66 ≈
10. 2,000 ÷ 66 ≈
11. 700 ÷ 67 ≈
12. 460 ÷ 67 ≈
13. 2,600 ÷ 66 ≈
14. 1,800 ÷ 66 ≈
15. 340 ÷ 67 ≈
16. 900 ÷ 67 ≈

Brain Builders

1. 24 ÷ 6.6 ≈
2. 1,600 ÷ 660 ≈
3. 72 ÷ 67 ≈
4. 8 ÷ 6.7
5. 480 ÷ 6.6 ≈
6. 52 ÷ 0.66 ≈
7. 3,200 ÷ 67 ≈
8. 54 ÷ 6.7 ≈
9. 380 ÷ 66 ≈
10. 62 ÷ 66 ≈

(See solutions on page 217)

Trick 53: Rapidly Estimate Division by 9 (or 0.9, 90, 900, etc.)

Strategy: Today's estimation techniques involve the approximate reciprocals 9 and 11. To estimate division by 9, **multiply by 11** and affix or insert any necessary zeroes or decimal point. You might want to review the "11-Trick" (Trick 8) on page 34 before studying the following examples.

Elementary Example #1

Estimate 34 ÷ 9

Step 1. Multiply: 34 × 11 = 374 (intermediary estimate).

Step 2. Apply T of R: A quick estimate puts the answer between 3 and 4.

Step 3. Insert a decimal point within the intermediary estimate, producing a final estimate of 3.74 (the true answer is about 3.78).

Thought Process Summary

$$34 \div 9 \quad \rightarrow \quad \begin{array}{r} 34 \\ \times 11 \\ \hline 374 \end{array} \quad \rightarrow \quad 3.74$$

Elementary Example #2

Estimate 78 ÷ 9

Step 1. Multiply: 78 × 11 = 858 (intermediary estimate).

Step 2. Apply T of R: A quick estimate puts the answer between 8 and 9.

Step 3. Insert a decimal point within the intermediary estimate, producing a final estimate of 8.58 (the true answer is about 8.67).

Thought Process Summary

$$78 \div 9 \rightarrow \begin{array}{r} 78 \\ \times 11 \\ \hline 858 \end{array} \rightarrow 8.58$$

Brain Builder #1

Estimate 4,300 ÷ 90

Step 1. Disregard the zeroes and think, "43 ÷ 9."

Step 2. Multiply: 43 × 11 = 473 (intermediary estimate).

Step 3. Apply T of R: A quick estimate puts the answer between 40 and 50.

Step 4. Insert a decimal point within the intermediary estimate, producing a final estimate of 47.3 (the true answer is about 47.78).

Thought Process Summary

$$4,300 \div 90 \rightarrow 43 \div 9 \rightarrow \begin{array}{r} 43 \\ \times 11 \\ \hline 473 \end{array} \rightarrow 47.3$$

Brain Builder #2

Estimate 53 ÷ 0.9

Step 1. Disregard the decimal point and think, "53 ÷ 9."

Step 2. Multiply: 53 × 11 = 583 (intermediary estimate).

Step 3. Apply T of R: A quick estimate puts the answer just above 53.

Step 4. Insert a decimal point within the intermediary estimate, producing a final estimate of 58.3 (the true answer is about 58.89).

Thought Process Summary

$$53 \div 0.9 \rightarrow 53 \div 9 \rightarrow \begin{array}{r} 53 \\ \times 11 \\ \hline 583 \end{array} \rightarrow 58.3$$

Number-Power Note: This trick will produce an estimate that is 1 percent *under* the true answer.

Elementary Exercises

Remember to multiply by 11 when working through these exercises.

1. $26 \div 9 \approx$
2. $71 \div 9 \approx$
3. $35 \div 9 \approx$
4. $62 \div 9 \approx$
5. $58 \div 9 \approx$
6. $47 \div 9 \approx$
7. $66 \div 9 \approx$
8. $95 \div 9 \approx$
9. $83 \div 9 \approx$
10. $49 \div 9 \approx$
11. $140 \div 9 \approx$
12. $220 \div 9 \approx$
13. $300 \div 9 \approx$
14. $700 \div 9 \approx$
15. $87 \div 9 \approx$
16. $110 \div 9 \approx$

Brain Builders

1. $380 \div 9 \approx$
2. $6,000 \div 90 \approx$
3. $1,700 \div 9 \approx$
4. $7.5 \div 0.9 \approx$
5. $410 \div 90 \approx$
6. $93 \div 90 \approx$
7. $1,000 \div 90 \approx$
8. $670 \div 9 \approx$
9. $5.5 \div 0.9 \approx$
10. $2,400 \div 90 \approx$

(See solutions on page 217)

Mathematical Curiosity #8

$$
\begin{aligned}
1 &= 1^2 \\
1+2+1 &= 2^2 \\
1+2+3+2+1 &= 3^2 \\
1+2+3+4+3+2+1 &= 4^2 \\
1+2+3+4+5+4+3+2+1 &= 5^2 \\
1+2+3+4+5+6+5+4+3+2+1 &= 6^2
\end{aligned}
$$

etc.

Trick 54: Rapidly Estimate Division by 11 (or 0.11, 1.1, 110, etc.)

Strategy: This estimation technique requires a bit more concentration than the previous one. To estimate division by 11, **multiply by 9** and affix or insert any necessary zeroes or decimal point. You might want to review the "9-Trick" (Trick 17) on page 60 before proceeding with the following examples.

Elementary Example #1

Estimate 40 ÷ 11

Step 1. Multiply: 40 × 9 = 360 (intermediary estimate).

Step 2. Apply T of R: A quick estimate puts the answer between 3 and 4.

Step 3. Insert a decimal point within the intermediary estimate, producing a final estimate of 3.6 (the true answer is about 3.64).

Thought Process Summary

$$40 \div 11 \rightarrow \begin{array}{r} 40 \\ \times 9 \\ \hline 360 \end{array} \rightarrow 3.6$$

Elementary Example #2

Estimate 90 ÷ 11

Step 1. Multiply: 90 × 9 = 810 (intermediary estimate).

Step 2. Apply T of R: A quick estimate puts the answer just above 8.

Step 3. Insert a decimal point within the intermediary estimate, producing a final estimate of 8.1 (the true answer is about 8.18).

Thought Process Summary

$$90 \div 11 \rightarrow \begin{array}{r} 90 \\ \times 9 \\ \hline 810 \end{array} \rightarrow 8.1$$

Brain Builder #1

Estimate 45 ÷ 1.1

Step 1. Disregard the decimal point and think, "45 ÷ 11."

Step 2. Multiply: 45 × 9 = 405 (intermediary estimate).

Step 3. Apply T of R: A quick estimate puts the answer at about 40.

Step 4. Insert a decimal point within the intermediary estimate, producing a final estimate of 40.5 (the true answer is about 40.9).

Thought Process Summary

$$45 \div 1.1 \rightarrow 45 \div 11 \rightarrow \begin{array}{r} 45 \\ \times 9 \\ \hline 405 \end{array} \rightarrow 40.5$$

Brain Builder #2

Estimate 750 ÷ 110

Step 1. Disregard the zeroes and think, "75 ÷ 11."

Step 2. Multiply: 75 × 9 = 675 (intermediary estimate).

Step 3. Apply T of R: A quick estimate puts the answer between 6 and 7.

Step 4. Insert a decimal point within the intermediary estimate, producing a final estimate of 6.75 (the true answer is about 6.82).

Thought Process Summary

$$750 \div 110 \rightarrow 75 \div 11 \rightarrow \begin{array}{r} 75 \\ \times 9 \\ \hline 675 \end{array} \rightarrow 6.75$$

Number-Power Note: This trick will produce an estimate that is 1 percent *under* the true answer.

Elementary Exercises

When doing these exercises, remember to multiply by 9.

1. 600 ÷ 11 ≈
2. 200 ÷ 11 ≈
3. 80 ÷ 11 ≈
4. 70 ÷ 11 ≈
5. 300 ÷ 11 ≈
6. 500 ÷ 11 ≈
7. 240 ÷ 11 ≈
8. 130 ÷ 11 ≈
9. 170 ÷ 11 ≈
10. 350 ÷ 11 ≈
11. 250 ÷ 11 ≈
12. 120 ÷ 11 ≈
13. 150 ÷ 11 ≈
14. 180 ÷ 11 ≈
15. 67 ÷ 11 ≈
16. 56 ÷ 11 ≈

Brain Builders

1. $16 \div 1.1 \approx$
2. $2,300 \div 110 \approx$
3. $340 \div 11 \approx$
4. $28 \div 1.1 \approx$
5. $7.8 \div 1.1 \approx$
6. $890 \div 11 \approx$
7. $3,600 \div 110 \approx$
8. $14 \div 1.1 \approx$
9. $4,000 \div 11 \approx$
10. $90 \div 110 \approx$

(See solutions on page 218)

Trick 55: Rapidly Estimate Division by 14 (or 0.14, 1.4, 140, etc.)

Strategy: Day 28 covers the last two estimation techniques presented. To estimate division by 14, **multiply by 7,** and affix or insert any necessary zeroes or decimal point. It's easy to remember this pair of approximate reciprocals because 7 is exactly half of 14. Let's look at a few examples.

Elementary Example #1

Estimate 60 ÷ 14

Step 1. Multiply: 60 × 7 = 420 (intermediary estimate).

Step 2. Apply T of R: A quick estimate puts the answer between 4 and 5.

Step 3. Insert a decimal point within the intermediary estimate, producing a final estimate of 4.2 (the true answer is about 4.29).

Thought Process Summary

$$60 \div 14 \rightarrow \begin{array}{r} 60 \\ \times 7 \\ \hline 420 \end{array} \rightarrow 4.2$$

Elementary Example #2

Estimate 110 ÷ 14

Step 1. Multiply: 110 × 7 = 770 (intermediary estimate).

Step 2. Apply T of R: A quick estimate puts the answer between 7 and 8.

Step 3. Insert a decimal point within the intermediary estimate, producing a final estimate of 7.7 (the true answer is about 7.86).

Thought Process Summary

$$110 \div 14 \rightarrow \begin{array}{r} 110 \\ \times 7 \\ \hline 770 \end{array} \rightarrow 7.7$$

Brain Builder #1

Estimate 300 ÷ 1.4

Step 1. Disregard the zeroes and decimal point and think, "3 ÷ 14."

Step 2. Multiply: 3 × 7 = 21 (intermediary estimate).

Step 3. Apply T of R: A quick estimate puts the answer at just above 200.

Step 4. Affix one zero to the intermediary estimate, producing a final estimate of 210 (the true answer is about 214.3).

Thought Process Summary

$$300 \div 1.4 \rightarrow 3 \div 14 \rightarrow \begin{array}{r} 3 \\ \times 7 \\ \hline 21 \end{array} \rightarrow 210$$

Brain Builder #2

Estimate 900 ÷ 140

Step 1. Disregard the zeroes and think, "9 ÷ 14."

Step 2. Multiply: 9 × 7 = 63 (intermediary estimate).

Step 3. Apply T of R: A quick estimate puts the answer between 6 and 7.

Step 4. Insert a decimal point within the intermediary estimate, producing a final estimate of 6.3 (the true answer is about 6.43).

Thought Process Summary

$$900 \div 140 \rightarrow 9 \div 14 \rightarrow \begin{array}{r} 9 \\ \times 7 \\ \hline 63 \end{array} \rightarrow 6.3$$

Number-Power Note: This trick will produce an estimate that is 2 percent *under* the true answer.

Elementary Exercises

Remember to multiply by 7 when working through these exercises.

1. $80 \div 14 \approx$
2. $90 \div 14 \approx$
3. $200 \div 14 \approx$
4. $500 \div 14 \approx$
5. $70 \div 14 \approx$
6. $400 \div 14 \approx$
7. $710 \div 14 \approx$
8. $150 \div 14 \approx$
9. $120 \div 14 \approx$
10. $210 \div 14 \approx$
11. $310 \div 14 \approx$
12. $130 \div 14 \approx$
13. $1,000 \div 14 \approx$
14. $91 \div 14 \approx$
15. $610 \div 14 \approx$
16. $300 \div 14 \approx$

Brain Builders

1. $510 \div 14 \approx$
2. $60 \div 1.4 \approx$
3. $1,100 \div 14 \approx$
4. $810 \div 140 \approx$
5. $8 \div 1.4 \approx$
6. $5,000 \div 140 \approx$
7. $4,100 \div 140 \approx$
8. $13 \div 1.4 \approx$
9. $1,200 \div 14 \approx$
10. $700 \div 1.4 \approx$

(See solutions on page 218)

Number Potpourri # 9

Here is evidence of the power of compound interest: See if you can guess approximately how much money would accumulate if you were to make an investment of $2,000 per year, every year, for 40 years (or a total of $80,000). Assume that your money is earning interest at 12 percent, compounded annually (that is, interest remains invested, to earn interest on itself).

(See answer on page 225)

Trick 56: Rapidly Estimate Division by 17 (or 0.17, 1.7, 170, etc.)

Strategy: Here is the tenth and last estimation technique for you to learn. To estimate division by 17, **multiply by 6**, and affix or insert any necessary zeroes or decimal point. This pair of approximate reciprocals can be remembered by noting that 6 is the difference between 1 and 7 (the digits of the number 17). Let's finish Week 4 by seeing how this trick works.

Elementary Example #1

Estimate 300 ÷ 17

Step 1. Disregard the zeroes and think, "3 ÷ 17."

Step 2. Multiply: 3 × 6 = 18 (intermediary estimate).

Step 3. Apply T of R: A quick estimate puts the answer in the teens. Therefore, the intermediary estimate of 18 is correct (the true answer is about 17.65).

Thought Process Summary

$$300 \div 17 \rightarrow 3 \div 17 \rightarrow \begin{array}{r} 3 \\ \times 6 \\ \hline 18 \end{array}$$

Elementary Example #2

Estimate 250 ÷ 17

Step 1. Disregard the zero and think, "25 ÷ 17."

Step 2. Multiply: 25 × 6 = 150 (intermediary estimate).

Step 3. Apply T of R: A quick estimate puts the answer in the teens.

Step 4. Insert a decimal point within the intermediary estimate, producing a final estimate of 15 (the true answer is about 14.7).

Thought Process Summary

$$250 \div 17 \rightarrow 25 \div 17 \rightarrow \begin{array}{r} 25 \\ \times 6 \\ \hline 150 \end{array} \rightarrow 15$$

Brain Builder #1

Estimate 80 ÷ 1.7

Step 1. Disregard the zero and decimal point and think, "8 ÷ 17."

Step 2. Multiply: 8 × 6 = 48 (intermediary estimate).

Step 3. Apply T of R: A quick estimate puts the answer near 50. The intermediary estimate of 48 is therefore correct (the true answer is about 47.06).

Thought Process Summary

$$80 \div 1.7 \rightarrow 8 \div 17 \rightarrow \begin{array}{r} 8 \\ \times 6 \\ \hline 48 \end{array}$$

Brain Builder #2

Estimate 6,500 ÷ 170

Step 1. Disregard the zeroes and think, "65 ÷ 17."

Step 2. Multiply: 65 × 6 = 390 (intermediary estimate).

Step 3 Apply T of R: A quick estimate puts the answer near 40.

Step 4. Insert a decimal point within the intermediary estimate, producing a final estimate of 39 (the true answer is about 38.2).

Thought Process Summary

$$6,500 \div 170 \rightarrow 65 \div 17 \rightarrow \begin{array}{r} 65 \\ \times 6 \\ \hline 390 \end{array} \rightarrow 39$$

Number-Power Note: This trick will produce an estimate that is 2 percent *over* the true answer. Trick 56 is the last estimation technique presented formally. However, if you'd like more, here's one last one: Division by 82, 83, 84, or 85 can be estimated simply by multiplying by 12.

Elementary Exercises

When doing these exercises, remember to multiply by 6.

1. 500 ÷ 17 ≈
2. 110 ÷ 17 ≈
3. 160 ÷ 17 ≈
4. 900 ÷ 17 ≈
5. 350 ÷ 17 ≈
6. 200 ÷ 17 ≈
7. 1,000 ÷ 17 ≈
8. 130 ÷ 17 ≈
9. 600 ÷ 17 ≈
10. 450 ÷ 17 ≈
11. 150 ÷ 17 ≈
12. 700 ÷ 17 ≈
13. 400 ÷ 17 ≈
14. 140 ÷ 17 ≈
15. 310 ÷ 17 ≈
16. 1,200 ÷ 17 ≈

Brain Builders

1. $81 \div 17 \approx$
2. $3,000 \div 17 \approx$
3. $180 \div 17 \approx$
4. $61 \div 1.7 \approx$
5. $7.1 \div 1.7 \approx$
6. $1,300 \div 170 \approx$
7. $25 \div 17 \approx$
8. $11 \div 1.7 \approx$
9. $6,000 \div 170 \approx$
10. $2,100 \div 17 \approx$

(See solutions on page 219)

Mathematical Curiosity #9

$$\begin{aligned}(0 \times 9) + 1 &= 1\\(1 \times 9) + 2 &= 11\\(12 \times 9) + 3 &= 111\\(123 \times 9) + 4 &= 1111\\(1234 \times 9) + 5 &= 11111\\(12345 \times 9) + 6 &= 111111\end{aligned}$$

etc.

Week 4 Quick Quiz

Let's see how many tricks from Week 4 you can remember and apply by taking this brief test. There's no time limit, but try to work through these items as rapidly as possible. Before you begin, just glance at the computations and try to identify the trick that you could use. When you flip ahead to the solutions, you will see which trick was intended. Remember that the symbol ≈ means "approximately equals."

Elementary Problems

1. $570 \div 15 =$
2. $72 \times 34 \approx$
3. $450 \div 75 =$
4. $49 \times 86 \approx$
5. $900 \div 66 \approx$
6. $76 \div 8 =$
7. $75 \times 56 =$
8. $150 \div 9 \approx$
9. $400 \div 11 \approx$
10. $360 \div 51 \approx$
11. $51 \times 64 \approx$
12. $160 \div 17 \approx$
13. $600 \div 14 \approx$
14. $900 \div 34 \approx$
15. $18 \times 67 \approx$
16. $33 \times 57 \approx$

Brain Builders

1. $6.4 \times 750 =$
2. $700 \div 4.9 \approx$
3. $3{,}800 \div 67 \approx$
4. $540 \times 3.3 \approx$
5. $12.8 \div 0.8 =$
6. $47 \div 1.1 \approx$
7. $1{,}300 \div 34 \approx$
8. $120 \div 1.4 \approx$
9. $6.6 \times 180 \approx$
10. $660 \div 7.5 =$
11. $1{,}200 \div 170 \approx$
12. $3{,}400 \div 90 \approx$
13. $36 \div 1.5 =$
14. $4.9 \times 112 \approx$

(See solutions on page 222)

Number Potpourri #10

The story goes that one of the questions on a high school science test was, "How would you measure the height of a building with the use of a barometer?" The teacher was hoping that her students would take into consideration the difference in air pressure between the top and bottom of the building, and apply the appropriate formula. However, one creative student was able to think of three ways: (1) go to the roof of the building, tie a rope around the barometer, lower the barometer to the street, and then measure the rope; (2) walk up the building's stairs, moving the barometer end-over-end on the wall, measure the length of the barometer, and perform the appropriate multiplication; or (3) drop the barometer from the roof of the building, see how long it takes to crash to the ground, and then apply the appropriate physics formula. When the student handed in his test, the teacher told him that his answers were not what she had in mind, and that he could have one more chance. So he went back to his desk and wrote, "Well, the only other way I can think of to answer this question is to tell the janitor that if he will measure the building for me, I will give him a barometer!"

Days 29 and 30: Grand Finale

Trick 57: Rapidly Multiply by Regrouping

Strategy: Days 29 and 30 involve some advanced tricks for multiplication and division. Trick 57 is similar to Trick 27, in that it deals with calculations that are better performed in two steps. For example, to calculate 43×6 in the conventional way would be a bit cumbersome. However, convert the problem to (40×6) plus (3×6), or 258, and it seems much easier. Another example is 32×15, which can be solved as $(30 \times 15) + (2 \times 15) = 480$. What you are doing, in effect, is multiplying from left to right. It takes a bit of practice and imagination to master this very practical technique.

Elementary Example #1
15 × 16

Regrouping: $(15 \times 15) + (15 \times 1) = 225 + 15 = 240$ (the answer).

Elementary Example #2
77 × 6

Regroup: $(70 \times 6) + (7 \times 6) = 420 + 42 = 462$ (the answer).

Brain Builder #1
23 × 35

Regroup: $(20 \times 35) + (3 \times 35) = 700 + 105 = 805$ (the answer).

Brain Builder #2
706 × 8

Regroup: $(700 \times 8) + (6 \times 8) = 5{,}600 + 48 = 5{,}648$ (the answer).

Number-Power Note: As stated above, it takes imagination to use this trick effectively. And speaking of imagination, see if you can figure out why it is faster to structure the problem 555×837 as 837×555, assuming that you are solving it in the conventional manner. Try it both ways, and you'll see why.

Elementary Exercises

If at first you don't succeed, regroup, and try again!

1. $57 \times 8 =$
2. $73 \times 5 =$
3. $7 \times 65 =$
4. $45 \times 46 =$
5. $84 \times 4 =$
6. $96 \times 3 =$
7. $6 \times 33 =$
8. $8 \times 22 =$
9. $64 \times 3 =$
10. $36 \times 7 =$
11. $5 \times 47 =$
12. $4 \times 56 =$
13. $8 \times 71 =$
14. $25 \times 26 =$
15. $6 \times 93 =$
16. $3 \times 87 =$

Brain Builders

1. $35 \times 37 =$
2. $31 \times 32 =$
3. $407 \times 6 =$
4. $708 \times 7 =$
5. $12 \times 35 =$
6. $21 \times 45 =$
7. $45 \times 22 =$
8. $60 \times 51 =$
9. $104 \times 30 =$
10. $106 \times 40 =$

(See solutions on page 219)

PARLOR TRICK #5 "I'D LIKE TO SPEAK TO THE WIZARD"

Have someone pick a playing card, or just think of one, and say it out loud for you and everyone else to hear. You then tell your guests that you know someone, known as "The Wizard," who can reveal the identity of the card by phone.

Pick up the phone and dial The Wizard's phone number. When someone answers, say "I'd like to speak to the wizard." Pause and say, "Hello, Wizard?" Pause again, and say, "There's someone who would like to speak to you." Give the phone to the person who chose the card, and the wizard will tell him or her exactly what card was chosen.

Prior to making the phone call, tell everyone exactly what you are going to say to the wizard, and do not vary these words. This way, your friends will realize that it isn't the wording that's giving away the identity of the card. If they are still suspicious after you've pulled off the trick, do it again with a different playing card, and they will see that the trick will work again with exactly the same wording.

At this point, it might be fun for you to take some time to figure out how the trick is done. If you can't figure it out, read on.

Strategy: As you may have figured out, the "wizard" is merely a co-conspirator, such as a friend or relative, who knows exactly what to do when he or she picks up the phone and hears the words, "I'd like to speak to the wizard."

First of all, if the person who answers the phone is not the "wizard," merely have that person put the wizard on the phone at this point. Then, the wizard will start to say over the phone, "Ace, two, three . . . " and so on, all the way up to "jack, queen, king." However, you will cut in and say, "Hello, Wizard?" just as the co-conspirator says the card number.

After you've said, "Hello, Wizard," the wizard will say, "Spades, clubs, hearts, diamonds." You, however, will cut in and say, "There's someone who would like to speak to you" just as the wizard has said the suit of the card chosen. The wizard will now know both the card number and the suit. Obviously, you are the only one who is allowed to listen to what the wizard is saying before he or she reveals the card to your guests.

This trick does take a little bit of practice (for example, you'll need to establish the appropriate pace on the part of the wizard), but it never fails to work and amuse, once you get the hang of it. You might wish to always have two or three "wizards" on call, so you can be almost guaranteed that one will be home when you want to make use of this excellent parlor trick.

Trick 58: Rapidly Multiply by Augmenting

Strategy: This trick is similar to Trick 57, in that you must use your imagination to apply it effectively. With this technique, however, you are going to pretend that one of the factors is slightly more than it is, and then adjust downward. For example, the problem 15×29 is the same as $(15 \times 30) - (15 \times 1) = 450 - 15 = 435$. Concentrate a wee bit harder to understand the following examples.

Elementary Example #1
7×28

Round up and subtract: $(7 \times 30) - (7 \times 2) = 210 - 14 = 196$ (the answer).

Elementary Example #2
38×6

Round up and subtract: $(40 \times 6) - (2 \times 6) = 240 - 12 = 228$ (the answer).

Brain Builder #1

29 × 30

Round up and subtract: $(30 \times 30) - (1 \times 30) = 900 - 30 = 870$ (the answer).

Brain Builder #2

75 × 98

Round up and subtract: $(75 \times 100) - (75 \times 2) = 7,500 - 150 = 7,350$ (the answer).

Number-Power Note: This technique is more difficult to apply than most, because you must perform rapid subtraction in your head. Nevertheless, it is a technique that will come in handy time and time again. Trick 17 ("Rapidly Multiply by 9") is also an example of an augmentation technique, in that you first multiply by 10 and then subtract.

Number Potpourri #11

How carefully do you pay attention to numbers? Let's see one last time. Approximately how high is Mt. Everest?

a. 4,500 feet
b. 16,000 feet
c. 29,000 feet
d. 41,000 feet

What is the approximate distance around the earth's equator?

a. 15,000 miles
b. 25,000 miles
c. 35,000 miles
d. 45,000 miles

(See answers on page 225)

Elementary Exercises

Pat yourself on the back if you can work through all these exercises quickly and accurately.

1. $19 \times 7 =$
2. $78 \times 4 =$
3. $6 \times 49 =$
4. $3 \times 68 =$
5. $98 \times 3 =$
6. $39 \times 5 =$
7. $7 \times 89 =$
8. $4 \times 58 =$
9. $29 \times 6 =$
10. $79 \times 2 =$
11. $4 \times 88 =$
12. $5 \times 98 =$
13. $99 \times 7 =$
14. $48 \times 6 =$
15. $5 \times 59 =$
16. $3 \times 69 =$

Brain Builders

1. $59 \times 60 =$
2. $29 \times 12 =$
3. $45 \times 98 =$
4. $70 \times 18 =$
5. $79 \times 15 =$
6. $39 \times 35 =$
7. $14 \times 48 =$
8. $45 \times 99 =$
9. $19 \times 25 =$
10. $68 \times 30 =$

(See solutions on page 219)

Trick 59: Rapidly Multiply Any Three-Digit or Larger Number by 11

Strategy: This trick is an advanced variation on Trick 8. To multiply a three-digit or larger number by 11, first carry down the ones digit of the number. Then add the ones digit to the tens digit, the tens digit to the hundreds digit, and so on, writing down in the answer space only the right-hand digit each time (until the last addition). It will be necessary to carry when the sum of any pair of digits exceeds 9. As you'll see below, this excellent trick is a lot easier to perform than it sounds.

Elementary Example #1

342 × 11

Step 1. Write 2 in the answer space as the ones digit.

Step 2. Add: 2 + 4 = 6 (tens-digit answer).

Step 3. Add: 4 + 3 = 7 (hundreds-digit answer).

Step 4. Write 3 in the answer space as the thousands digit.

Step 5. Summarize: The answer is 3,762

Thought Process Summary

3 4 2		3 4 (2)		3 (4 2)		(3 4) 2		(3) 4 2
×1 1	→	×1 1	→	×1 1	→	×1 1	→	×1 1
		(2)		(6) 2		(7) 6 2		(3), 7 6 2

Elementary Example #2

726 × 11

Step 1. Write 6 in the answer space as the ones digit.

Step 2. Add: 6 + 2 = 8 (tens-digit answer).

Step 3. Add: 2 + 7 = 9 (hundreds-digit answer).

Step 4. Write 7 in the answer space as the thousands digit.

Step 5. Summarize: The answer is 7,986.

Thought Process Summary

7 2 6		7 2 6		7 2 6		7 2 6		7 2 6
×1 1	→	×1 1	→	×1 1	→	×1 1	→	×1 1
		6		8 6		9 8 6		7, 9 8 6

Brain Builder #1

594 × 11

Step 1. Write 4 in the answer space as the ones digit.

Step 2. Add: 4 + 9 = 13 (3 is the tens-digit answer; carry the 1).

Step 3. Add: 9 + 5 + 1 carried = 15 (5 is the hundreds-digit answer, carry the 1).

Step 4. Add: 5 + 1 carried = 6 (thousands-digit answer).

Step 5. Summarize: The answer is 6,534.

Thought Process Summary

				1		1 1		1 1
5 9 4		5 9 4		5 9 4		5 9 4		5 9 4
×1 1	→	×1 1	→	×1 1	→	×1 1	→	×1 1
		4		3 4		5 3 4		6, 5 3 4

Brain Builder #2

8,097 × 11

Step 1. Write 7 in the answer space as the ones digit.

Step 2. Add: 7 + 9 = 16 (6 is the tens-digit answer; carry the 1).

Step 3. Add: 9 + 0 + 1 carried = 10 (0 is the hundreds digit answer; carry the 1).

Step 4. Add: 0 + 8 + 1 carried = 9 (thousands-digit answer).

Step 5. Write 8 in the answer space as the ten-thousands-digit answer.

Step 6. Summarize: The answer is 89,067.

Thought Process Summary

		1	1 1	1 1	1 1
8,097	8,097	8,097	8,097	8,097	8,097
×11 →	×11 →	×11 →	×11 →	×11 →	×11
	7	67	067	9,067	89,067

Number-Power Note: Although this trick will also work with decimal points and affixed zeroes, they have been omitted to simplify the explanation.

Elementary Exercises

Impress yourself by completing these exercises without showing any work.

1. 321 × 11 =
2. 143 × 11 =
3. 524 × 11 =
4. 613 × 11 =
5. 11 × 435 =
6. 11 × 354 =
7. 11 × 807 =
8. 11 × 262 =
9. 172 × 11 =
10. 452 × 11 =

Brain Builders

1. 577 × 11 =
2. 369 × 11 =
3. 475 × 11 =
4. 934 × 11 =
5. 11 × 669 =
6. 11 × 395 =
7. 11 × 555 =
8. 11 × 857 =
9. 3,141 × 11 =
10. 2,680 × 11 =

(See solutions on page 220)

Mathematical Curiosity #10

5,363,222,357 × 2,071,723 = 11,111,111,111,111,111

Trick 60: Rapidly Divide by 9, 99, 999, etc.

Strategy: We conclude our thirty-day program with the magical number 9. Dividing by all 9s will produce a repeating pattern, as you will see shortly. In each case, however, the numerator must be smaller than the denominator. For example, when dividing by 99, the numerator cannot exceed 98. Take special note of how the pattern repeats in the five examples that follow.

Elementary Example #1

4 ÷ 9 = 0.444444...

Elementary Example #2

62 ÷ 99 = 0.626262...

Brain Builder #1

8 ÷ 999 = 0.008008008...

Brain Builder #2

31 ÷ 999 = 0.031031031...

Brain Builder #3

409 ÷ 999 = 0.409409409...

Number-Power Note: When the numerator *exceeds* the denominator, a repeating decimal pattern will still appear, but not in the same way it does in the examples above. For example, 472 ÷ 99 = 4.767676...

Elementary Exercises

Carry these calculations to six decimal places. When you finish, proceed to the Pre-test Revisited, Final Exam, and Conclusion.

1. 5 ÷ 9 =
2. 2 ÷ 9 =
3. 3 ÷ 99 =
4. 1 ÷ 99 =
5. 8 ÷ 99 =
6. 6 ÷ 99 =
7. 76 ÷ 99 =
8. 53 ÷ 99 =
9. 50 ÷ 99 =
10. 94 ÷ 99 =
11. 25 ÷ 99 =
12. 47 ÷ 99 =
13. 88 ÷ 99 –
14. 61 ÷ 99 =
15. 7 ÷ 99 =
16. 4 ÷ 99 =

Brain Builders

1. 2 ÷ 999 =
2. 9 ÷ 999 =
3. 70 ÷ 999 =
4. 39 ÷ 999 =
5. 12 ÷ 999 =
6. 84 ÷ 999 =
7. 26 ÷ 999 =
8. 71 ÷ 999 =
9. 514 ÷ 999 =
10. 763 ÷ 999 =

(See solutions on page 220)

Pre-Test Revisited

How Many of These Calculations Can You Do in Two Minutes?

Directions: Work each calculation below as quickly as you can, without using a calculator. Your time limit is two minutes, and you may do the problems in any order. When your time is up, refer back to page 7 and check your answers. You won't believe how much you've learned since you originally took this pre-test.

1. $99 \times 85 =$
2. $700 \div 25 =$
3. $3.5 \times 110 =$
4. $4,600 \div 50 =$
5. $1.9 \times 210 =$
6. $425 - 387 =$
7. $31 \times 31 =$
8. $7 + 24 + 343 + 50 =$
9. $22 \times 18 =$
10. $31.5 \div 3.5 =$
11. $1 + 2 + 3 + 4 + 5 + 6 + 7 + 8 + 9 + 10 =$
12. $120 \div 1.5 =$
13. $65 \times 65 =$
14. $74 \times 101 =$
15. $163 - 128 =$
16. $109 \times 104 =$

Final Exam

Well, this is it! Take a deep breath and relax. When you finish, grade your own test (be honest now) and then go and celebrate. Use any rapid-math trick you like, unless otherwise indicated. Ready, set, go!

Elementary Problems

1. $93 \times 6 =$
2. $7 \times 89 =$
3. $312 \times 11 =$
4. Carry to six decimal places: $83 \div 99 =$
5. $46 + 17 + 5 + 29 + 33 + 2 =$
6. $7 + 2 + 1 + 5 + 8 + 5 =$
7. $85 \times 4 =$
8. $81 \div 5 =$
9. $35^2 =$
10. $11 \times 44 =$
11. $25 \times 68 =$
12. $78 \times 99 =$
13. $81 \times 79 =$
14. $1,500 \div 125 =$
15. $9 \times 56 =$
16. $21 \times 29 =$
17. $4.5 \times 16 =$
18. $53^2 =$
19. $71^2 =$
20. $115 - 88 =$
21. $25 + 82 + 30 + 8 =$
22. $1 + 2 + 3 + \ldots + 15 =$

23. $420 \div 75 =$
24. Using the "casting out nines" method, indicate whether this calculation is probably correct or definitely incorrect:
$92 + 42 = 133$
25. $33 \times 27 \approx$
26. $640 \div 49 \approx$
27. $710 \div 9 \approx$
28. $66 \times 24 \approx$
29. $200 \div 14 \approx$
30. $500 \div 11 \approx$

Brain Builders

1. $980 \div 40 =$
2. $47 \times 50 =$
3. $5.7 \times 101 =$
4. $91 \div 2.5 =$
5. $1.25 \times 96 =$
6. Using the "casting out nines" method, indicate whether this calculation is probably correct or definitely incorrect:
$384 \times 761 = 292,224$
7. $1.2 \times 230 =$
8. $\$180 \times 15\% =$
9. $180 \div 4.5 =$
10. $4.1^2 =$
11. $88 \times 9.2 =$
12. $14 \times 22 =$
13. $154 - 78 =$
14. $103 \times 112 =$
15. Add these without carrying:
```
 726
 149
 335
 608
 527
+924
----
```
16. Add in two sections:
```
 247
 918
 336
 157
 444
 790
 688
+805
----
```
17. $93 - 57 =$
18. $32 \times 750 =$
19. $104 \div 80 =$
20. $5,700 \div 150 =$
21. $350 \div 33 \approx$
22. $510 \times 2.4 \approx$
23. $4,800 \div 67 \approx$
24. $210 \div 1.7 \approx$
25. Carry to six decimal places:
$314 \div 999 =$
26. $11 \times 837 =$
27. $49 \times 13 =$
28. $45 \times 21 =$
29. $115^2 =$
30. $9.9 \times 9.9 =$

Congratulations!

(See solutions on page 223)

— In Conclusion —

Don't feel obligated to use every number-power trick illustrated in this book. Even if you are thoroughly engrossed in the material, it is unlikely you will remember all 60 tricks at any given time. Moreover, you will find that some tricks are more useful to you than others.

Accordingly, I recommend that you take some time to determine which of the 60 tricks you find the most useful, most interesting, and easiest to remember. Then focus on those tricks in your daily life, and don't worry so much about the others. Every now and then, re-examine all 60 to modify your working repertoire.

As I pointed out earlier, the more tricks you commit to memory, the greater the likelihood that you will be able to apply number power for a given situation. However, if you try to use more tricks than you can handle comfortably, you may become a bit frustrated and overwhelmed. The real trick is for you to determine the optimal number you can use efficiently. Maybe that number is 10 or maybe it is 40. Experimentation and time will tell.

I have two final recommendations. First, don't use, for example, only multiplication tricks or only tricks for addition. Have a balanced repertoire of addition, subtraction, multiplication, division, and estimation techniques. Second, if possible, learn the advanced applications (brain builders) for each trick, not just the basic (elementary) applications. This way, you open yourself up to a far greater world of possibilities.

Making Number Power a Part of Your Daily Life

Now that your mind is a veritable arsenal of rapid calculation techniques, you must practice them so that they will remain fresh in your mind. If you are a teacher or have a math-related position, you will automatically use number power every day. However, for everyone else, let me suggest the following practical applications:

- When you go to the supermarket, estimate the total cost of groceries before they are rung up at the checkstand. Your estimate should be within 5 percent of the actual total.
- The next time you fill your gas tank, reset your trip odometer (if your car has one) to zero. Then, when you refill your tank the next time, use number power to calculate the gas mileage your car has achieved since the last fill-up (divide gallons purchased into miles driven).
- When you buy something at a department store, use cash and perform mental math to calculate the exact amount of change you should receive.

You may be surprised to find that you are being short-changed (or perhaps are receiving too much change) on a relatively frequent basis.

- The next time you are about to pay the check at a restaurant, don't panic or shuffle through your wallet or purse to find your 15-percent gratuity table. You will feel enormous power upon calculating the tip in a mere few seconds.
- Back at the supermarket, use number power to comparison shop. That is, calculate the price per unit of one product versus another, or of the regular size versus the jumbo economy size. You may be surprised to learn that occasionally the jumbo size is more expensive on a per-unit basis than the regular size. If your supermarket already discloses unit-price information, don't cheat by referring to it.
- When traveling, make note of numbers on license plates, billboards, and buildings, and manipulate those numbers in some way. For example, if you see a license plate that reads "645W327," mentally add 327 to 645, and then perform subtraction on the two numbers. This will keep your mind "mathematically fit."
- This application is more a problem-solving strategy than a rapid-calculation technique. Nevertheless, it is an easy method that few people think to apply. The next time you purchase an item that is marked, say, 30 percent off, don't waste your time multiplying the original price by 30 percent, and then subtracting the discount from the original price. Instead, simply multiply the original price by 70 percent (100 percent minus 30 percent) to arrive, in one step, at the sale price.
- Now see if you can think of an application that you think would apply to your daily activities. Use your imagination—the list of possibilities is endless!

I hope you enjoyed learning number power as much as I enjoyed teaching it. But remember that anything that has been learned will easily be forgotten if not put to use on a regular basis. On the other hand, anything that has been learned well can be relearned quickly at a later time.

For those interested, I hold rapid-math workshops on a fairly regular basis. If you would like me to address your organization, school, or business, please contact me at P.O. Box 4145, Chatsworth, CA 91313-4145, for more information.

I would also enjoy receiving your comments, suggestions, or perhaps your very own rapid-math techniques. Please write to me at the above address. I look forward to hearing from you.

Thank you, and good luck!

Edward Julius

— Solutions —

Exercises

Trick 1

1. 280
2. 4,800
3. 15,000
4. 54,000
5. 8,400
6. 2,250
7. 20,000
8. 9,600
9. 4
10. 800
11. 40
12. 50
13. 700
14. 20

Trick 2

1. 24
2. 920
3. 350
4. 750
5. 78
6. 144
7. 54
8. 9.3
9. 600
10. 300
11. 300
12. 4,000
13. 90
14. 300

Trick 3

Elementary

1. 140
2. 92
3. 56
4. 340
5. 164
6. 104
7. 220
8. 288
9. 244
10. 68
11. 380
12. 192
13. 116
14. 332
15. 260
16. 212

Brain Builders

1. 2,160
2. 144
3. 30
4. 60
5. 364
6. 316
7. 2,280
8. 176
9. 100
10. 392

Trick 4

Elementary

1. 12
2. 17
3. 45
4. 33
5. 65
6. 24
7. 14
8. 22
9. 35
10. 55
11. 16
12. 18
13. 95
14. 85
15. 105
16. 13

Brain Builders

1. 11
2. 9
3. 28
4. 23.5
5. 13.5
6. 44
7. 3.6
8. 58
9. 20.25
10. 2.45

Trick 5

Elementary

1. 80
2. 190
3. 440
4. 210
5. 370
6. 290
7. 110
8. 380
9. 310
10. 140
11. 330
12. 470
13. 270
14. 410
15. 480
16. 220

Brain Builders

1. 425
2. 245
3. 165
4. 485
5. 2,750
6. 6.1
7. 395
8. 2,550
9. 105
10. 3,400

Trick 6

Elementary

1. 5.4
2. 10.6
3. 14.4
4. 13.4
5. 23.6
6. 19
7. 8.2
8. 9.8
9. 24.4
10. 2.8
11. 15.2
12. 16.2
13. 6.6
14. 11.6
15. 19.6
16. 12.8

Brain Builders

1. 4.6
2. 3.7
3. 16.6
4. 6.66
5. 3.8
6. 0.84
7. 87
8. 18.4
9. 1.22
10. 0.92

Trick 7

Elementary

1. 1,225
2. 7,225
3. 9,025
4. 625
5. 3,025
6. 5,625
7. 2,025
8. 225
9. 4,225
10. 9,025
11. 7,225
12. 1,225
13. 625
14. 3,025
15. 5,625
16. 2,025

Brain Builders

1. 11,025
2. 1,225
3. 562,500
4. 72.25
5. 422.5
6. 2,250
7. 132.25
8. 3,025
9. 0.2025
10. 9,025

Trick 8

Elementary

1. 682
2. 198
3. 385
4. 891
5. 286
6. 484
7. 638
8. 1,012
9. 187
10. 759
11. 341
12. 814
13. 1,056
14. 429
15. 517
16. 1,089

Brain Builders

1. 297
2. 71.5
3. 9.13
4. 616
5. 143
6. 36,300
7. 858
8. 2.42
9. 10,340
10. 539

Trick 9

Elementary

1. 300
2. 1,100
3. 1,300
4. 400
5. 1,600
6. 2,200
7. 600
8. 1,400
9. 2,300
10. 1,700
11. 1,200
12. 2,000
13. 800
14. 2,100
15. 1,800
16. 2,400

Brain Builders

1. 850
2. 1,950
3. 1,450
4. 350
5. 185
6. 7,000
7. 215
8. 150
9. 210
10. 4,500

Trick 10

Elementary

1. 3.2
2. 12
3. 7.2
4. 96
5. 12.8
6. 26
7. 48
8. 34
9. 3.8
10. 16
11. 360
12. 60
13. 20
14. 88
15. 30
16. 10.8

Brain Builders

1. 140
2. 4.44
3. 0.92
4. 22
5. 6.8
6. 16.8
7. 8.88
8. 328
9. 5.2
10. 13.2

Trick 11

Elementary

1. 5,940
2. 7,425
3. 891
4. 8,712
5. 3,465
6. 6,039
7. 6,534
8. 4,752
9. 7,920
10. 2,178
11. 396
12. 5,346
13. 8,217
14. 3,861
15. 9,603
16. 1,089

Brain Builders

1. 5,148
2. 900.9
3. 76.23
4. 257.4
5. 1,980
6. 326,700
7. 712.8
8. 435.6
9. 5.643
10. 2,970

Trick 12

Elementary

1. 1,515
2. 6,262
3. 3,939
4. 808
5. 9,393
6. 4,141
7. 8,787
8. 7,070
9. 1,212
10. 4,545
11. 8,181
12. 2,323
13. 606
14. 7,878
15. 3,232
16. 9,999

Brain Builders

1. 4.848
2. 63.63
3. 363.6
4. 85.85
5. 11,110
6. 929.2
7. 303
8. 1,919
9. 898.9
10. 515.1

Trick 13

Elementary

1. 195
2. 255
3. 899
4. 224
5. 1,224
6. 6,399
7. 9,999
8. 3,024
9. 143
10. 168
11. 399
12. 288
13. 3,599
14. 624
15. 323
16. 7,224

Brain Builders

1. 39.9
2. 562.4
3. 143
4. 2.55
5. 202.4
6. 999.9
7. 16,800
8. 24.99
9. 32.3
10. 4,224

Trick 14

Elementary

1. Definitely incorrect
2. Probably correct
3. Probably correct
4. Definitely incorrect
5. Definitely incorrect
6. Probably correct
7. Definitely incorrect
8. Probably correct
9. Probably correct
10. Probably correct
11. Definitely incorrect
12. Probably correct
13. Definitely incorrect
14. Probably correct
15. Definitely incorrect
16. Probably correct

Brain Builders

1. Probably correct
2. Definitely incorrect
3. Probably correct
4. Probably correct
5. Definitely incorrect
6. Definitely incorrect
7. Probably correct
8. Probably correct
9. Probably correct
10. Definitely incorrect
11. Probably correct
12. Probably correct

Trick 15

Elementary

1. 2,000
2. 5,000
3. 11,000
4. 7,000
5. 3,000
6. 8,000
7. 12,000
8. 4,000
9. 10,000
10. 20,000
11. 9,000
12. 6,000
13. 15,000
14. 13,000
15. 1,000
16. 30,000

Brain Builders

1. 2,500
2. 7,500
3. 4,500
4. 3,500
5. 80
6. 40
7. 1,100
8. 120
9. 700
10. 600

Trick 16

Elementary

1. 1.6
2. 48
3. 8.8
4. 6.4
5. 32
6. 72
7. 80
8. 5.6
9. 24
10. 40
11. 9.6
12. 8.88
13. 4.8
14. 12
15. 2
16. 16

Brain Builders

1. 32
2. 7.2
3. 6
4. 0.888
5. 1.6
6. 0.48
7. 2.8
8. 56
9. 4
10. 2

Trick 17

Elementary

1. 117
2. 216
3. 315
4. 153
5. 108
6. 225
7. 495
8. 135
9. 252
10. 603
11. 306
12. 144
13. 504
14. 162
15. 171
16. 207

Brain Builders

1. 70.2
2. 8,010
3. 243
4. 261
5. 342
6. 855
7. 423
8. 62.1
9. 32,400
10. 522

Trick 18

Elementary

1. 540
2. 216
3. 192
4. 900
5. 264
6. 168
7. 180
8. 660
9. 384
10. 252
11. 204
12. 420
13. 288
14. 1,020
15. 372
16. 276

Brain Builders

1. 408
2. 204
3. 66
4. 2,760
5. 114
6. 180
7. 3.84
8. 2,280
9. 26.4
10. 102

Trick 19

Elementary

1. $1.20
2. $2.40
3. 660
4. 390
5. $2.10
6. $5.70
7. 510
8. 630
9. $2.70
10. $3.30
11. 930
12. 810
13. $3.60
14. $8.70
15. 690
16. 780

Brain Builders

1. 1,320
2. 1,080
3. $.84
4. $.72
5. 5,400
6. 240
7. $.96
8. $10.20
9. 270
10. 375

Trick 20

Elementary

1. 216
2. 1,221
3. 7,224
4. 609
5. 3,016
6. 9,021
7. 624
8. 7,209
9. 3,024
10. 5,621
11. 1,209
12. 4,224
13. 5,609
14. 9,009
15. 2,016
16. 7,216

Brain Builders

1. 2,024
2. 2,090
3. 56.24
4. 300.9
5. 616
6. 901.6
7. 1,224
8. 2,210
9. 42.21
10. 0.9024

Trick 21

Elementary

1. 21
2. 165
3. 693
4. 76
5. 78
6. 51
7. 45
8. 84
9. 195
10. 72
11. 253
12. 52
13. 81
14. 98
15. 57
16. 85

Brain Builders

1. 510
2. 840
3. 79.2
4. 520
5. 720
6. 760
7. 540
8. 550
9. 720
10. 680

Trick 22

Elementary

1. 6
2. 4
3. 9
4. 8
5. 5
6. 7
7. 6
8. 4
9. 8
10. 6
11. 4
12. 20
13. 30
14. 8
15. 3
16. 14

Brain Builders

1. 14
2. 70
3. 30
4. 7
5. 6
6. 5
7. 7
8. 40
9. 30
10. 7

Trick 23

Elementary

1. 2,704
2. 3,136
3. 3,481
4. 2,809
5. 3,249
6. 3,025
7. 2,916
8. 3,364
9. 2,601
10. 3,481
11. 3,136
12. 2,704
13. 2,809
14. 3,249
15. 3,025
16. 2,601

Brain Builders

1. 32.49
2. 260.1
3. 302.5
4. 270.4
5. 348.1
6. 313.6
7. 33.64
8. 28,090
9. 291,600
10. 348.1

Trick 24

Elementary

1. 961
2. 3,721
3. 8,281
4. 121
5. 5,041
6. 1,681
7. 6,561
8. 2,601
9. 441
10. 961
11. 3,721
12. 5,041
13. 1,681
14. 8,281
15. 2,601
16. 6,561

Brain Builders

1. 441
2. 26.01
3. 504.1
4. 828.1
5. 96,100
6. 372.1
7. 1,681
8. 65.61
9. 260.1
10. 4,410

Trick 25

Elementary

1. 273
2. 308
3. 480
4. 1,764
5. 1,683
6. 325
7. 484
8. 1,802
9. 728
10. 1,935
11. 1,435
12. 1,265
13. 810
14. 1,071
15. 1,156
16. 540

Brain Builders

1. 504
2. 2,904
3. 4,104
4. 8,004
5. 5,544
6. 4,930
7. 3,528
8. 1,377
9. 3,168
10. 3,078

Trick 26

Elementary

1. 165
2. 221
3. 957
4. 96
5. 117
6. 192
7. 896
8. 2,021
9. 621
10. 437
11. 252
12. 1,677
13. 3,021
14. 9,996
15. 4,896
16. 2,597

Brain Builders

1. 1,170
2. 285
3. 357
4. 35.96
5. 4,221
6. 43.7
7. 809.6
8. 252
9. 56.21
10. 655.7

Trick 27

Elementary

1. 98
2. 108
3. 128
4. 108
5. 330
6. 90
7. 288
8. 286
9. 144
10. 126
11. 144
12. 374
13. 126
14. 96
15. 112
16. 80

Brain Builders

1. 9.6
2. 144
3. 3.74
4. 1,120
5. 126
6. 14.4
7. 1,080
8. 12.6
9. 286
10. 1,280

Trick 28

Elementary

1. 10,201
2. 11,235
3. 10,918
4. 11,881
5. 10,608
6. 11,556
7. 11,009
8. 11,236
9. 10,920
10. 10,807
11. 11,772
12. 10,605
13. 11,021
14. 11,448
15. 11,445
16. 11,016

Brain Builders

1. 11,872
2. 11,960
3. 12,091
4. 12,099
5. 12,875
6. 13,566
7. 12,096
8. 12,285
9. 16,867
10. 12,688

Trick 29

Elementary

1. 14
2. 17
3. 28
4. 25
5. 51
6. 26
7. 24
8. 53
9. 36
10. 19
11. 48
12. 37
13. 18
14. 37
15. 22
16. 45

Brain Builders

1. 25
2. 34
3. 43
4. 36
5. 58
6. 45
7. 59
8. 91
9. 75
10. 27

Trick 30

Elementary

1. 16
2. 45
3. 58
4. 56
5. 42
6. 39
7. 43
8. 53
9. 88
10. 57
11. 66
12. 62
13. 93
14. 17
15. 59
16. 87

Brain Builders

1. 27
2. 56
3. 26
4. 53
5. 33
6. 48
7. 77
8. 36
9. 49
10. 66

Trick 31

Elementary

1. 25
2. 34
3. 37
4. 32
5. 18
6. 42
7. 35
8. 13
9. 27
10. 12
11. 33
12. 21
13. 38
14. 41
15. 28
16. 34

Brain Builders

1. 55
2. 38
3. 18
4. 29
5. 49
6. 49
7. 39
8. 51
9. 75
10. 32

Trick 32

Elementary

1. 74
2. 84
3. 122
4. 95
5. 134
6. 132
7. 112
8. 74
9. 163
10. 113
11. 105
12. 133
13. 125
14. 122
15. 122
16. 113

Brain Builders

1. 243
2. 176
3. 183
4. 234
5. 194
6. 173
7. 171
8. 263
9. 223
10. 134

Trick 33

Elementary

1. 44
2. 37
3. 50
4. 33
5. 43
6. 35
7. 42
8. 34
9. 50
10. 38
11. 45
12. 42
13. 35
14. 44
15. 45
16. 41

Brain Builders

1. 45
2. 36
3. 39
4. 43
5. 50
6. 36
7. 39
8. 37
9. 37
10. 49

Trick 34

Elementary

1. 451
2. 454
3. 397
4. 404
5. 453
6. 447
7. 445
8. 420
9. 392
10. 427
11. 546
12. 425
13. 436
14. 483
15. 455
16. 395

Brain Builders

1. 4,688
2. 4,239
3. 4,276
4. 4,306
5. 4,191
6. 4,374
7. 4,683
8. 4,882
9. 10,011
10. 10,163

Trick 35

Elementary

1. 135
2. 129
3. 131
4. 135
5. 108
6. 176
7. 120
8. 139
9. 107
10. 135

Brain Builders

1. 205
2. 177
3. 236
4. 205
5. 176
6. 191
7. 211
8. 204
9. 189
10. 138

Trick 36

Elementary

1. 128
2. 123
3. 131
4. 140
5. 105
6. 175
7. 122
8. 139
9. 117
10. 136
11. 124
12. 149
13. 130
14. 122
15. 135
16. 105

Brain Builders

1. 203
2. 175
3. 234
4. 203
5. 174
6. 189
7. 209
8. 212
9. 192
10. 138

Trick 37

Elementary

1. 43
2. 528
3. 449
4. 499
5. 499
6. 522
7. 533
8. 471
9. 415
10. 505
11. 601
12. 446
13. 532
14. 555
15. 492
16. 448

Brain Builders

1. 5,007
2. 4,877
3. 5,060
4. 4,971
5. 4,781
6. 4,982
7. 4,936
8. 5,417
9. 5,133
10. 4,892

Trick 38

Elementary

1. 588
2. 524
3. 642
4. 622
5. 549
6. 548
7. 518
8. 559
9. 492
10. 509
11. 519
12. 544
13. 546
14. 564
15. 547
16. 571

Brain Builders

1. 5,615
2. 5,470
3. 5,465
4. 5,721
5. 5,199
6. 5,757
7. 5,541
8. 6,045
9. 55,969
10. 52,785

Trick 39

Elementary

1. 114
2. 127
3. 127
4. 142
5. 106
6. 142
7. 128
8. 135
9. 144
10. 141
11. 126
12. 129
13. 153
14. 146
15. 158
16. 137

Brain Builders

1. 214
2. 229
3. 200
4. 220
5. 269
6. 242
7. 276
8. 233
9. 254
10. 277

Trick 40

Elementary

1. 15
2. 21
3. 36
4. 120
5. 78
6. 210
7. 153
8. 105
9. 325
10. 253
11. 171
12. 378
13. 231
14. 45
15. 465
16. 66

Brain Builders

1. 1,275
2. 630
3. 3,003
4. 4,095
5. 1,378
6. 3,321
7. 2,485
8. 1,711
9. 4,465
10. 2,016

Trick 41

Elementary

1. 24
2. 33
3. 22
4. 41
5. 66
6. 45
7. 37
8. 78
9. 28
10. 29
11. 73
12. 37
13. 34
14. 72
15. 61
16. 135

Brain Builders

1. 35
2. 48
3. 56
4. 49
5. 24
6. 73
7. 82
8. 47
9. 46
10. 63

Trick 42

Elementary

1. Probably correct
2. Definitely incorrect
3. Definitely incorrect
4. Probably correct
5. Probably correct
6. Definitely incorrect
7. Probably correct
8. Probably correct
9. Probably correct
10. Definitely incorrect
11. Probably correct
12. Probably correct
13. Definitely incorrect
14. Probably correct
15. Probably correct
16. Definitely incorrect

Brain Builders

1. Definitely incorrect
2. Probably correct
3. Probably correct
4. Definitely incorrect
5. Probably correct
6. Definitely incorrect
7. Probably correct
8. Definitely incorrect
9. Probably correct
10. Definitely incorrect
11. Probably correct
12. Probably correct
13. Definitely incorrect
14. Probably correct

Trick 43

Elementary

1. 2,100
2. 2,400
3. 3,300
4. 1,200
5. 900
6. 3,900
7. 4,200
8. 600
9. 3,600
10. 6,300

Brain Builders

1. 4,200
2. 24,000
3. 1,200
4. 63
5. 1,650
6. 1,950
7. 2,250
8. 1,350
9. 1,050
10. 39

Trick 44

Elementary

1. 4.8
2. 20
3. 36
4. 4.4
5. 5.6
6. 28
7. 24
8. 8
9. 8.8
10. 52
11. 60
12. 0.4
13. 6.8
14. 16
15. 132
16. 6.4

Brain Builders

1. 124
2. 9.6
3. 11.6
4. 92
5. 24
6. 0.6
7. 0.88
8. 72
9. 1.28
10. 0.84

Trick 45

Elementary

1. 4.5
2. 6.5
3. 5.5
4. 6.25
5. 7.5
6. 17.5
7. 8.5
8. 3.75
9. 10.5
10. 22.5
11. 12.5
12. 32.5
13. 15
14. 9.5
15. 11.5
16. 35

Brain Builders

1. 55
2. 42.5
3. 1.15
4. 25
5. 0.95
6. 6.25
7. 3.75
8. 13
9. 4.5
10. 15

Trick 46

Elementary

1. 36
2. 28
3. 6.6
4. 24
5. 14
6. 56
7. 42
8. 18
9. 8
10. 64
11. 32
12. 52
13. 22
14. 46
15. 62
16. 38

Brain Builders

1. 54
2. 4.4
3. 1.6
4. 5.8
5. 34
6. 0.6
7. 30
8. 6.2
9. 2.4
10. 0.52

Trick 47

Elementary

1. 500 (true answer = 495)
2. 2,200 (true answer = 2,178)
3. 1,600 (true answer = 1,632)
4. 3,200 (true answer = 3,264)
5. 1,900 (true answer = 1,881)
6. 700 (true answer = 693)
7. 2,400 (true answer = 2,448)
8. 2,800 (true answer = 2,856)
9. 3,100 (true answer = 3,069)
10. 1,500 (true answer = 1,485)
11. 900 (true answer = 918)
12. 1,800 (true answer = 1,836)
13. 2,600 (true answer = 2,574)
14. 1,100 (true answer = 1,089)
15. 2,500 (true answer = 2,550)
16. 2,300 (true answer = 2,346)

Brain Builders

1. 290 (true answer = 287.1)
2. 130 (true answer = 128.7)
3. 400 (true answer = 408)
4. 330 (true answer = 336.6)
5. 800 (true answer = 792)
6. 210 (true answer = 207.9)
7. 1,800 (true answer = 1,836)
8. 27 (true answer = 27.54)
9. 200 (true answer = 198)
10. 600 (true answer = 594)

Trick 48

Elementary

1. 21 (true answer = 21.2121 . . .)
2. 3.3 (true answer = 3.333 . . .)
3. 7.5 (true answer ≈ 7.35)
4. 24 (true answer ≈ 23.53)
5. 4.8 (true answer = 4.8484 . . .)
6. 6.9 (true answer = 6.9696 . . .)
7. 12 (true answer ≈ 11.76)
8. 4.2 (true answer ≈ 4.12)
9. 9.3 (true answer = 9.3939 . . .)
10. 15 (true answer = 15.1515 . . .)
11. 22.5 (true answer ≈ 22.06)
12. 5.4 (true answer ≈ 5.29)
13. 27 (true answer = 27.2727 . . .)
14. 6.6 (true answer = 6.666 . . .)
15. 8.4 (true answer ≈ 8.24)
16. 18 (true answer ≈ 17.65)

Brain Builders

1. 6 (true answer = 6.0606 . . .)
2. 5.1 (true answer = 5.1515 . . .)
3. 12.9 (true answer ≈ 12.65)
4. 24.6 (true answer ≈ 24.12)
5. 108 (true answer = 109.0909 . . .)
6. 3.9 (true answer = 3.9393 . . .)
7. 8.1 (true answer ≈ 7.94)
8. 13.5 (true answer ≈ 13.24)
9. 72 (true answer = 72.7272 . . .)
10. 27.3 (true answer = 27.5757 . . .)

Trick 49

Elementary

1. 3,200 (true answer = 3,136)
2. 1,100 (true answer = 1,078)
3. 3,900 (true answer = 3,978)
4. 2,800 (true answer = 2,856)
5. 1,450 (true answer = 1,421)
6. 650 (true answer = 637)
7. 2,250 (true answer = 2,295)
8. 1,550 (true answer = 1,581)
9. 2,800 (true answer = 2,744)
10. 3,500 (true answer = 3,430)
11. 4,600 (true answer = 4,692)
12. 4,300 (true answer = 4,386)
13. 2,650 (true answer = 2,597)
14. 1,850 (true answer = 1,813)
15. 4,900 (true answer = 4,998)
16. 3,550 (true answer = 3,621)

Brain Builders

1. 80 (true answer = 78.4)
2. 375 (true answer = 367.5)
3. 12 (true answer = 12.24)
4. 4,850 (true answer = 4,947)
5. 440 (true answer = 431.2)
6. 550 (true answer = 539)
7. 335 (true answer = 341.7)
8. 41.5 (true answer = 42.33)
9. 90 (true answer = 88.2)
10. 355 (true answer = 347.9)

Trick 50

Elementary

1. 4.8 (true answer ≈ 4.9)
2. 1.82 (true answer ≈ 1.86)
3. 1.76 (true answer ≈ 1.73)
4. 9 (true answer ≈ 8.82)
5. 11.4 (true answer ≈ 11.63)
6. 6.4 (true answer ≈ 6.53)
7. 1.58 (true answer ≈ 1.55)
8. 12.2 (true answer ≈ 11.96)
9. 3.6 (true answer ≈ 3.67)
10. 1.88 (true answer ≈ 1.92)
11. 14.6 (true answer ≈ 14.31)
12. 4.2 (true answer ≈ 4.12)
13. 6 (true answer ≈ 6.12)
14. 7.2 (true answer ≈ 7.35)
15. 15.4 (true answer ≈ 15.1)
16. 12.8 (true answer ≈ 12.55)

Brain Builders

1. 8.6 (true answer ≈ 8.78)
2. 1.94 (true answer ≈ 1.98)
3. 30 (true answer ≈ 29.41)
4. 14 (true answer ≈ 13.73)
5. 16.4 (true answer ≈ 16.73)
6. 54 (true answer ≈ 55.1)
7. 11.6 (true answer ≈ 11.37)
8. 6.8 (true answer = 6.666...)
9. 150 (true answer ≈ 153.06)
10. 19.8 (true answer ≈ 20.2)

Trick 51

Elementary

1. 800 (true answer = 792)
2. 3,000 (true answer = 2,970)
3. 2,400 (true answer = 2,412)
4. 1,400 (true answer = 1,407)
5. 6,000 (true answer = 5,940)
6. 1,200 (true answer = 1,188)
7. 5,000 (true answer = 5,025)
8. 2,200 (true answer = 2,211)
9. 3,400 (true answer = 3,366)
10. 6,200 (true answer = 6,138)
11. 4,000 (true answer = 4,020)
12. 1,600 (true answer = 1,608)
13. 1,800 (true answer = 1,782)
14. 5,600 (true answer = 5,544)
15. 4,200 (true answer = 4,221)
16. 6,400 (true answer = 6,432)

Brain Builders

1. 320 (true answer = 316.8)
2. 1,000 (true answer = 990)
3. 58 (true answer = 58.29)
4. 280 (true answer = 281.4)
5. 2,000 (true answer = 1,980)
6. 5,400 (true answer = 5,346)
7. 480 (true answer = 482.4)
8. 380 (true answer = 381.9)
9. 520 (true answer = 514.8)
10. 660 (true answer = 653.4)

Trick 52

Elementary

1. 2.1 (true answer = 2.1212...)
2. 3.3 (true answer = 3.333...)
3. 9 (true answer ≈ 8.96)
4. 5.4 (true answer ≈ 5.37)
5. 4.2 (true answer = 4.2424...)
6. 6.3 (true answer = 6.3636...)
7. 9.6 (true answer ≈ 9.55)
8. 8.7 (true answer ≈ 8.66)
9. 18 (true answer = 18.1818...)
10. 30 (true answer = 30.3030...)
11. 10.5 (true answer ≈ 10.45)
12. 6.9 (true answer ≈ 6.87)
13. 39 (true answer = 39.3939...)
14. 27 (true answer = 27.2727...)
15. 5.1 (true answer ≈ 5.07)
16. 13.5 (true answer ≈ 13.43)

Brain Builders

1. 3.6 (true answer = 3.6363...)
2. 2.4 (true answer = 2.4242...)
3. 1.08 (true answer ≈ 1.075)
4. 1.2 (true answer ≈ 1.194)
5. 72 (true answer = 72.7272...)
6. 78 (true answer = 78.7878...)
7. 48 (true answer ≈ 47.76)
8. 8.1 (true answer ≈ 8.06)
9. 5.7 (true answer = 5.7575...)
10. 0.93 (true answer = 0.9393...)

Trick 53

Elementary

1. 2.86 (true answer = 2.888...)
2. 7.81 (true answer = 7.888...)
3. 3.85 (true answer = 3.888...)
4. 6.82 (true answer = 6.888...)
5. 6.38 (true answer = 6.444...)
6. 5.17 (true answer = 5.222...)
7. 7.26 (true answer = 7.333...)
8. 10.45 (true answer = 10.555...)
9. 9.13 (true answer = 9.222...)
10. 5.39 (true answer = 5.444...)
11. 15.4 (true answer = 15.555...)
12. 24.2 (true answer = 24.444...)
13. 33 (true answer = 33.333...)
14. 77 (true answer = 77.777...)
15. 9.57 (true answer = 9.666...)
16. 12.1 (true answer = 12.222...)

Brain Builders

1. 41.8 (true answer = 42.222...)
2. 66 (true answer = 66.666...)
3. 187 (true answer = 188.888...)
4. 8.25 (true answer = 8.333...)
5. 4.51 (true answer = 4.555...)
6. 1.023 (true answer = 1.0333...)
7. 11 (true answer = 11.111...)
8. 73.7 (true answer = 74.444...)
9. 6.05 (true answer = 6.111...)
10. 26.4 (true answer = 26.666...)

Trick 54

Elementary

1. 54 (true answer = 54.5454...)
2. 18 (true answer = 18.1818...)
3. 7.2 (true answer = 7.2727...)
4. 6.3 (true answer = 6.3636...)
5. 27 (true answer = 27.2727...)
6. 45 (true answer = 45.4545...)
7. 21.6 (true answer = 21.8181...)
8. 11.7 (true answer = 11.8181...)
9. 15.3 (true answer = 15.4545...)
10. 31.5 (true answer = 31.8181...)
11. 22.5 (true answer = 22.7272...)
12. 10.8 (true answer = 10.9090...)
13. 13.5 (true answer = 13.6363...)
14. 16.2 (true answer = 16.3636...)
15. 6.03 (true answer = 6.0909...)
16. 5.04 (true answer = 5.0909...)

Brain Builders

1. 14.4 (true answer = 14.5454...)
2. 20.7 (true answer = 20.9090...)
3. 30.6 (true answer = 30.9090...)
4. 25.2 (true answer = 25.4545...)
5. 7.02 (true answer = 7.0909...)
6. 80.1 (true answer = 80.9090...)
7. 32.4 (true answer = 32.7272...)
8. 12.6 (true answer = 12.7272...)
9. 360 (true answer = 363.6363...)
10. 0.81 (true answer = 0.8181...)

Trick 55

Elementary

1. 5.6 (true answer ≈ 5.71)
2. 6.3 (true answer ≈ 6.43)
3. 14 (true answer ≈ 14.29)
4. 35 (true answer ≈ 35.71)
5. 4.9 (true answer = 5)
6. 28 (true answer ≈ 28.57)
7. 49.7 (true answer ≈ 50.71)
8. 10.5 (true answer ≈ 10.71)
9. 8.4 (true answer ≈ 8.57)
10. 14.7 (true answer = 15)
11. 21.7 (true answer ≈ 22.14)
12. 9.1 (true answer ≈ 9.29)
13. 70 (true answer ≈ 71.43)
14. 6.37 (true answer = 6.5)
15. 42.7 (true answer ≈ 43.57)
16. 21 (true answer ≈ 21.43)

Brain Builders

1. 35.7 (true answer ≈ 36.43)
2. 42 (true answer ≈ 42.86)
3. 77 (true answer ≈ 78.57)
4. 5.67 (true answer ≈ 5.79)
5. 5.6 (true answer ≈ 5.71)
6. 35 (true answer ≈ 35.71)
7. 28.7 (true answer ≈ 29.29)
8. 9.1 (true answer ≈ 9.29)
9. 84 (true answer ≈ 85.71)
10. 490 (true answer = 500)

Trick 56

Elementary

1. 30 (true answer ≈ 29.41)
2. 6.6 (true answer ≈ 6.47)
3. 9.6 (true answer ≈ 9.41)
4. 54 (true answer ≈ 52.94)
5. 21 (true answer ≈ 20.59)
6. 12 (true answer ≈ 11.76)
7. 60 (true answer ≈ 58.82)
8. 7.8 (true answer ≈ 7.65)
9. 36 (true answer ≈ 35.29)
10. 27 (true answer ≈ 26.47)
11. 9 (true answer ≈ 8.82)
12. 42 (true answer ≈ 41.18)
13. 24 (true answer ≈ 23.53)
14. 8.4 (true answer ≈ 8.24)
15. 18.6 (true answer ≈ 18.24)
16. 72 (true answer ≈ 70.59)

Brain Builders

1. 4.86 (true answer ≈ 4.76)
2. 180 (true answer ≈ 176.47)
3. 10.8 (true answer ≈ 10.59)
4. 36.6 (true answer ≈ 35.88)
5. 4.26 (true answer ≈ 4.18)
6. 7.8 (true answer ≈ 7.65)
7. 1.5 (true answer ≈ 1.47)
8. 6.6 (true answer ≈ 6.47)
9. 36 (true answer ≈ 35.29)
10. 126 (true answer ≈ 123.53)

Trick 57

Elementary

1. 456
2. 365
3. 455
4. 2,070
5. 336
6. 288
7. 198
8. 176
9. 192
10. 252
11. 235
12. 224
13. 568
14. 650
15. 558
16. 261

Brain Builders

1. 1,295
2. 992
3. 2,442
4. 4,956
5. 420
6. 945
7. 990
8. 3,060
9. 3,120
10. 4,240

Trick 58

Elementary

1. 133
2. 312
3. 294
4. 204
5. 294
6. 195
7. 623
8. 232
9. 174
10. 158
11. 352
12. 490
13. 693
14. 288
15. 295
16. 207

Brain Builders

1. 3,540
2. 348
3. 4,410
4. 1,260
5. 1,185
6. 1,365
7. 672
8. 4,455
9. 475
10. 2,040

Trick 59

Elementary

1. 3,531
2. 1,573
3. 5,764
4. 6,743
5. 4,785
6. 3,894
7. 8,877
8. 2,882
9. 1,892
10. 4,972

Brain Builders

1. 6,347
2. 4,059
3. 5,225
4. 10,274
5. 7,359
6. 4,345
7. 6,105
8. 9,427
9. 34,551
10. 29,480

Trick 60

Elementary

1. 0.555555...
2. 0.222222...
3. 0.030303...
4. 0.010101...
5. 0.080808...
6. 0.060606...
7. 0.767676...
8. 0.535353...
9. 0.505050...
10. 0.949494...
11. 0.252525...
12. 0.474747...
13. 0.888888...
14. 0.616161...
15. 0.070707...
16. 0.040404...

Brain Builders

1. 0.002002...
2. 0.009009...
3. 0.070070...
4. 0.039039...
5. 0.012012...
6. 0.084084...
7. 0.026026...
8. 0.071071...
9. 0.514514...
10. 0.763763...

Quick Quizzes

Week 1

Elementary

1. 180 (Trick 3)
2. 8.8 (Trick 6)
3. 396 (Trick 8)
4. 30 (Trick 1)
5. 54,000 (Trick 1)
6. 1,800 (Trick 9)
7. 6,435 (Trick 11)
8. 4,225 (Trick 7)
9. 899 (Trick 13)
10. 19 (Trick 4)
11. 260 (Trick 5)
12. 9,393 (Trick 12)
13. 210 (Tricks 1 and 2)
14. 400 (Tricks 1 and 2)
15. 28 (Trick 10)
16. Definitely incorrect (Trick 14)

Brain Builders

1. 2,350 (Trick 5)
2. 1,225 (Trick 7)
3. 638 (Trick 8)
4. 14.5 (Trick 4)
5. 1.42 (Trick 6)
6. 84.4 (Trick 10)
7. 2,080 (Trick 3)
8. 159.9 (Trick 13)
9. 470 (Trick 5)
10. 562.5 (Trick 7)
11. 90 (Trick 9)
12. 373.7 (Trick 12)
13. 8.415 (Trick 11)
14. Probably correct (Trick 14)

Week 2

Elementary

1. 234 (Trick 17)
2. 4,221 (Trick 20 or 26)
3. 63 (Trick 21)
4. 126 (Trick 27)
5. 7,000 (Trick 15)
6. 2,704 (Trick 23)
7. 744 (Trick 25)
8. 780 (Trick 18)
9. 609 (Trick 20)
10. 8 (Trick 22)
11. 2,496 (Trick 26)
12. 961 (Trick 24)
13. $5.40 (Trick 19)
14. 56 (Trick 16)
15. 3,364 (Trick 23)
16. 11,124 (Trick 28)

Brain Builders

1. 216 (Trick 18 or 20)
2. 24 (Trick 16)
3. 3,721 (Trick 24)
4. 56.16 (Trick 20)
5. 43.7 (Trick 26 and 24)
6. 40 (Trick 22)
7. 800 (Trick 15)
8. 12,272 (Trick 28)
9. 324.9 (Trick 23)
10. 9,600 (Trick 19)
11. 112 (Trick 27)
12. 324 (Trick 17)
13. 720 (Trick 21)
14. 3,901 (Trick 25)

Week 3

Elementary

1. 132 (Trick 32)
2. 38 (Trick 29 or 30)
3. 300 (Trick 37)
4. 396 (Trick 38)
5. 36 (Trick 29 or 31)
6. 210 (Trick 40)
7. 122 (Trick 32)
8. 38 (Trick 33)
9. 145 (Trick 39)
10. Definitely incorrect (Trick 42)
11. 38 (Trick 29 or 30)
12. 68 (Trick 29, 30, 31, or 41)
13. 27 (Trick 29 or 31)
14. 52 (Trick 29)
15. 130 (Trick 35 or 36)
16. 348 (Trick 34)

Brain Builders

1. 58 (Trick 29 or 30)
2. 245 (Trick 32)
3. 40 (Trick 33)
4. 5,151 (Trick 40)
5. 325 (Trick 30 or 31)
6. 53 (Trick 29, 30, or 41)
7. 158 (Trick 35 or 36)
8. 134 (Trick 35 or 36)
9. 264 (Trick 39)
10. 81 (Trick 30 or 31)
11. 12,808 (Trick 38)
12. Probably correct (Trick 42)
13. 3,031 (Trick 34)
14. 4,061 (Trick 37)

Week 4

Elementary

1. 38 (Trick 46)
2. Approximately 2,400 (Trick 47)
3. 6 (Trick 44)
4. Approximately 4,300 (Trick 49)
5. Approximately 13.5 (Trick 52)
6. 9.5 (Trick 45)
7. 4,200 (Trick 43)
8. Approximately 16.5 (Trick 53)
9. Approximately 36 (Trick 54)
10. Approximately 7.2 (Trick 50)
11. Approximately 3,200 (Trick 49)
12. Approximately 9.6 (Trick 56)
13. Approximately 42 (Trick 55)
14. Approximately 27 (Trick 48)
15. Approximately 1,200 (Trick 51)
16. Approximately 1,900 (Trick 47)

Brain Builders

1. 4,800 (Trick 43)
2. Approximately 140 (Trick 50)
3. Approximately 57 (Trick 52)
4. Approximately 1,800 (Trick 47)
5. 16 (Trick 45)
6. Approximately 42.3 (Trick 54)
7. Approximately 39 (Trick 48)
8. Approximately 84 (Trick 55)
9. Approximately 1,200 (Trick 51)
10. 88 (Trick 44)
11. Approximately 7.2 (Trick 56)
12. Approximately 37.4 (Trick 53)
13. 24 (Trick 46)
14. Approximately 560 (Trick 49)

Solutions to Final Exam

Elementary Problems

1. 558 (Trick 57)
2. 623 (Trick 58)
3. 3,432 (Trick 59)
4. 0.838383 . . . (Trick 60)
5. 132 (Trick 35 or 36)
6. 28 (Trick 33)
7. 340 (Trick 3)
8. 16.2 (Trick 6)
9. 1,225 (Trick 7)
10. 484 (Trick 8)
11. 1,700 (Trick 9)
12. 7,722 (Trick 11)
13. 6,399 (Trick 13)
14. 12 (Trick 16)
15. 504 (Trick 17)
16. 609 (Trick 20)
17. 72 (Trick 21)
18. 2,809 (Trick 23)
19. 5,041 (Trick 25)
20. 27 (Trick 30 or 31)
21. 145 (Trick 39)
22. 120 (Trick 40)
23. 5.6 (Trick 44)
24. Definitely incorrect (Trick 42)
25. Approximately 900 (Trick 47)
26. Approximately 12.8 (Trick 50)
27. Approximately 78.1 (Trick 53)
28. Approximately 1,600 (Trick 51)
29. Approximately 14 (Trick 55)
30. Approximately 45 (Trick 54)

Brain Builders

1. 24.5 (Trick 4)
2. 2,350 (Trick 5)
3. 575.7 (Trick 12)
4. 36.4 (Trick 10)
5. 120 (Trick 15)
6. Probably correct (Trick 14)
7. 276 (Trick 18)
8. 27 (Trick 19)
9. 40 (Trick 22)
10. 16.81 (Trick 24)
11. 809.6 (Trick 26)
12. 308 (Trick 27)
13. 76 (Trick 29)
14. 11,536 (Trick 28)
15. 3,269 (Trick 34)
16. 4,385 (Trick 38)
17. 36 (Trick 29 or 41)
18. 24,000 (Trick 43)
19. 1.3 (Trick 45)
20. 38 (Trick 46)
21. Approximately 10.5 (Trick 48)
22. Approximately 1,200 (Trick 49)
23. Approximately 72 (Trick 52)
24. Approximately 126 (Trick 56)
25. 0.314314 . . . (Trick 60)
26. 9,207 (Trick 59)
27. 637 (Trick 58)
28. 945 (Trick 57)
29. 13,225 (Trick 7)
30. 98.01 (Trick 11)

Parlor Tricks

Parlor Trick 1

A. 6
B. 61
C. 80
D. 54

E. 29
F. 13
G. 95
H. 78

I. 42
J. 38
K. 7
L. 99

Parlor Trick 2

A. Wednesday
B. Tuesday
C. Wednesday
D. Monday

E. Saturday
F. Thursday
G. Friday
H. Monday

I. Sunday
J. Thursday
K. Tuesday
L. Wednesday

Parlor Trick 3

A. 56,100
B. 64,260
C. 36,720

D. 155,142
E. 488,892
F. 65,534

Parlor Trick 4

A. 13
B. 51
C. 80

D. 125
E. 20
F. 180

Number Potpourris

3. 4,100 (not 5,000, as the majority of people obtain)
4. 100 times more powerful
5. It means that chances are 1 in 3 (not even odds, as many people think).
6. 250 million (United States), 5.4 billion (world)
7. Trick question: *all* of them are divisible by 8. However, only 152 is *evenly* divisible by 8.
8. Fifty-proof liquor contains 25 percent alcohol.
9. It would accumulate to over $1.5 million!
11. Mt. Everest is 29,000 feet high; the earth's circumference is 25,000 miles.

Summary of the 60 Number-Power Tricks for Handy Reference

TRICK 1: When multiplying or dividing, initially disregard any affixed zeroes. Then reaffix, if necessary, upon completing calculation.

TRICK 2: When multiplying or dividing, initially disregard any decimal points. Then reinsert, if necessary, upon completing calculation.

TRICK 3: To multiply a number by 4, double the number, and then double once again.

TRICK 4: To divide a number by 4, halve the number, and then halve once again.

TRICK 5: To multiply a number by 5, divide the number by 2.

TRICK 6: To divide a number by 5, multiply the number by 2.

TRICK 7: To square a number that ends in 5, multiply the tens digit by the next whole number, and then affix the number 25.

TRICK 8: To multiply a two-digit number by 11, add the digits of the number, and insert the sum within the number itself. Carry if necessary.

TRICK 9: To multiply a number by 25, divide the number by 4.

TRICK 10: To divide a number by 25, multiply the number by 4.

TRICK 11: To multiply a one- or two-digit number by 99, subtract 1 from the number, and affix the difference between 100 and the number.

TRICK 12: To multiply a one- or two-digit number by 101, write the number twice; if a one-digit number, insert a zero in the middle.

TRICK 13: To multiply two numbers whose difference is 2, square the number in the middle and subtract 1.

TRICK 14: To check multiplication, multiply digit sums of factors and compare with digit sum of product. To check division, treat as multiplication, and check in same manner.

TRICK 15: To multiply a number by 125, divide the number by 8.

TRICK 16: To divide a number by 125, multiply the number by 8.

TRICK 17: To multiply a number by 9, multiply the number by 10 and then subtract the number.

TRICK 18: To multiply a number by 12, multiply the number by 10 and add twice the number itself.

TRICK 19: To multiply a number by 15, multiply the number by 10 and add half of the product.

TRICK 20: To multiply two 2-digit numbers whose tens digits are the same and whose ones digits add up to ten, multiply the tens digit by the next whole number, and affix the product of the ones digits.

TRICK 21: To multiply a number by 1.5, 2.5, or the like, first halve the number and double the 1.5 (or 2.5, or the like.) Then multiply.

TRICK 22: To divide a number by 1.5, 2.5, or the like, first double both the number and the 1.5 (or 2.5, or the like.) Then divide.

TRICK 23: To square a two-digit number beginning in 5, add 25 to the ones digit and affix the square of the ones digit.

TRICK 24: To square a two-digit number ending in 1, compute as in the following example: $31^2 = 30^2 + 30 + 31 = 961$

TRICK 25: To multiply two 2-digit numbers without showing work, first multiply the ones digits together, then "cross-multiply," and finally multiply the tens digits together. Carry if necessary.

TRICK 26: To multiply two numbers whose difference is 4, square the number exactly in the middle and subtract 4.

TRICK 27: When a calculation seems slightly beyond your reach, divide one number into two smaller ones. For example, view 8×14 as $8 \times 7 \times 2$.

TRICK 28: To multiply two numbers that are just over 100, begin the answer with a 1. Then affix first the sum, and then the product, of the ones digits.

TRICK 29: To subtract rapidly, view subtraction as addition, and work from left to right.

TRICK 30: To subtract when numbers are on opposite sides of 100, 200, or the like, determine the two "distances" and add.

TRICK 31: To make subtraction easier, alter the minuend and subtrahend in the same direction to make the subtrahend a multiple of 10, 100, or the like.

TRICK 32: To make addition easier, round up an addend, add the addends, and then subtract the number that was added when rounding.

TRICK 33: Group numbers to be added in combinations of 10, add numbers slightly out of order, and "see" two or three numbers as their sum.

TRICK 34: When adding columns of numbers, enter the column totals without carrying, moving one column to the left each time.

TRICK 35: To mentally add a column of numbers, add one number at a time—first the tens digit, then the ones digit.

TRICK 36: To mentally add a column of numbers, first add all the tens digits, then add all the ones digits.

TRICK 37: When adding long columns of numbers, lightly cross out a digit every time you exceed 9, and proceed with just the ones digit.

TRICK 38: When adding long columns of numbers, divide the column into smaller, more manageable sections.

TRICK 39: When adding just a few numbers, it is fastest to begin with the largest number and to end with the smallest.

TRICK 40: To add $1 + 2 + 3 + \ldots + n$, multiply n by $(n + 1)$, and then divide by 2.

TRICK 41: If you prefer not to "subtract by adding," then you can subtract in two steps—first the tens digits, then the ones.

TRICK 42: To check addition, add digit sums of addends, and compare with digit sum of the answer. To check subtraction, treat as addition and check in the same manner.

TRICK 43: To multiply a number by 75, multiply the number by $\frac{3}{4}$.

TRICK 44: To divide a number by 75, multiply the number by $1\frac{1}{3}$.

TRICK 45: To divide a number by 8, multiply the number by $1\frac{1}{4}$.

TRICK 46: To divide a number by 15, multiply the number by $\frac{2}{3}$.

TRICK 47: To estimate multiplication by 33 or 34, divide by 3.

TRICK 48: To estimate division by 33 or 34, multiply by 3.

TRICK 49: To estimate multiplication by 49 or 51, divide by 2.

TRICK 50: To estimate division by 49 or 51, multiply by 2.

TRICK 51: To estimate multiplication by 66 or 67, multiply by $\frac{2}{3}$.

TRICK 52: To estimate division by 66 or 67, multiply by 1.5.

TRICK 53: To estimate division by 9, multiply by 11.

TRICK 54: To estimate division by 11, multiply by 9.

TRICK 55: To estimate division by 14, multiply by 7.

TRICK 56: To estimate division by 17, multiply by 6.

TRICK 57: When a multiplication seems slightly beyond your grasp, regroup, as in the following example: $43 \times 6 = (40 \times 6) + (3 \times 6) = 258$.

TRICK 58: When a multiplicand or multiplier is just shy of a multiple of 10 or 100, round up and subtract, as in the following example: $15 \times 29 = (15 \times 30) - (15 \times 1) = 435$.

TRICK 59: To multiply a three-digit or larger number by 11, first carry down the ones digit of the number. Then add the ones and tens digits, the tens and hundreds digit, and so forth. Carry when necessary.

TRICK 60: Dividing by all nines produces a repeating pattern. For example, $2 \div 9 = 0.222\ldots$, $37 \div 99 = 0.3737\ldots$, and $486 \div 999 = 0.486486\ldots$